Studies in Computational Intelligence

Volume 674

Series editor

Janusz Kacprzyk, Polish Academy of Sciences, Warsaw, Poland
e-mail: kacprzyk@ibspan.waw.pl

About this Series

The series "Studies in Computational Intelligence" (SCI) publishes new developments and advances in the various areas of computational intelligence—quickly and with a high quality. The intent is to cover the theory, applications, and design methods of computational intelligence, as embedded in the fields of engineering, computer science, physics and life sciences, as well as the methodologies behind them. The series contains monographs, lecture notes and edited volumes in computational intelligence spanning the areas of neural networks, connectionist systems, genetic algorithms, evolutionary computation, artificial intelligence, cellular automata, self-organizing systems, soft computing, fuzzy systems, and hybrid intelligent systems. Of particular value to both the contributors and the readership are the short publication timeframe and the worldwide distribution, which enable both wide and rapid dissemination of research output.

More information about this series at http://www.springer.com/series/7092

Katsuhide Fujita · Quan Bai
Takayuki Ito · Minjie Zhang
Fenghui Ren · Reyhan Aydoğan
Rafik Hadfi
Editors

Modern Approaches to Agent-based Complex Automated Negotiation

Editors
Katsuhide Fujita
Tokyo University of Agriculture and Technology
Tokyo
Japan

Quan Bai
School of Engineering, Computer and Mathematical Sciences
Auckland University of Technology
Auckland
New Zealand

Takayuki Ito
School of Techno-Business Administration
Nagoya Institute of Technology
Nagoya
Japan

Minjie Zhang
School of Computing and Information Technology
University of Wollongong
Wollongong, NSW
Australia

Fenghui Ren
School of Computing and Information Technology
University of Wollongong
Wollongong, NSW
Australia

Reyhan Aydoğan
Ozyegin University
Istanbul
Turkey

Rafik Hadfi
Department of Computer Science and Engineering, Graduate School of Engineering
Nagoya Institute of Technology
Nagoya
Japan

ISSN 1860-949X ISSN 1860-9503 (electronic)
Studies in Computational Intelligence
ISBN 978-3-319-84684-2 ISBN 978-3-319-51563-2 (eBook)
DOI 10.1007/978-3-319-51563-2

© Springer International Publishing AG 2017
Softcover reprint of the hardcover 1st edition 2017
This work is subject to copyright. All rights are reserved by the Publisher, whether the whole or part of the material is concerned, specifically the rights of translation, reprinting, reuse of illustrations, recitation, broadcasting, reproduction on microfilms or in any other physical way, and transmission or information storage and retrieval, electronic adaptation, computer software, or by similar or dissimilar methodology now known or hereafter developed.
The use of general descriptive names, registered names, trademarks, service marks, etc. in this publication does not imply, even in the absence of a specific statement, that such names are exempt from the relevant protective laws and regulations and therefore free for general use.
The publisher, the authors and the editors are safe to assume that the advice and information in this book are believed to be true and accurate at the date of publication. Neither the publisher nor the authors or the editors give a warranty, express or implied, with respect to the material contained herein or for any errors or omissions that may have been made. The publisher remains neutral with regard to jurisdictional claims in published maps and institutional affiliations.

Printed on acid-free paper

This Springer imprint is published by Springer Nature
The registered company is Springer International Publishing AG
The registered company address is: Gewerbestrasse 11, 6330 Cham, Switzerland

Preface

Complex automated negotiations are a widely studied, emerging area in the field of autonomous agents and multi-agent systems. In general, automated negotiations can be complex, since there are many factors that characterize such negotiations. These factors include the number of issues, dependency between issues, representation of the utility, negotiation protocol, negotiation form (bilateral or multi-party), and time constraints. Software agents can support automation or simulation of such complex negotiations on behalf of their owners and can provide them with adequate bargaining strategies. In many multi-issue bargaining settings, negotiation becomes more than a zero-sum game, so bargaining agents have an incentive to cooperate in order to achieve efficient win-win agreements. Also, in a complex negotiation, there could be multiple issues that are interdependent. Thus, an agent's utility will become more complex than simple utility functions. Further, negotiation forms and protocols could be different between bilateral situations and multi-party situations. To realize such a complex automated negotiation, we have to incorporate advanced artificial intelligence technologies including search, CSP, graphical utility models, Bayesian nets, auctions, utility graphs, and predicting and learning methods. Applications could include e-commerce tools, decision-making support tools, negotiation support tools, collaboration tools, and others. For this book, we solicited papers on all aspects of such complex automated negotiations that are studied in the field of autonomous agents and multi-agent systems.

This book includes Part I: Agent-Based Complex Automated Negotiations, and Part II: Automated Negotiation Agents Competition. Each chapter in Part I is an extended version of an International Workshop on Agent-based Complex Automated Negotiations (ACAN'15) paper after peer reviews by three PC members. Part II includes Automated Negotiating Agents Competition (ANAC'15), in which automated agents who have different negotiation strategies and are implemented by different developers automatically negotiate in several negotiation domains. ANAC is an international competition in which automated negotiation strategies, submitted by several universities across the world, are evaluated in a tournament style. The purpose of the competition is to steer the research in the area of bilateral multi-issue, closed negotiation. Closed negotiation, when opponents do not reveal their preferences to each other, is an important class of real-life negotiations. Negotiating agents designed using a heuristic approach need extensive evaluation, typically through simulations and empirical analysis, since it is usually impossible to predict precisely how the system and the constituent agents will behave in a wide variety of circumstances, using purely theoretical tools. This book includes rules, results, agents, and domain descriptions for ANAC2015 submitted by organizers and finalists. The reports from the ANAC2015 competition highlight the important aspects that should be considered in future works on automated negotiation.

Finally, we would like to extend our sincere thanks to all authors. This book would not have been possible without the valuable support and contributions of those who cooperated with us.

Tokyo, Japan — Katsuhide Fujita
Auckland, New Zealand — Quan Bai
Nagoya, Japan — Takayuki Ito
Wollongong, Australia — Minjie Zhang
Wollongong, Australia — Fenghui Ren
Istanbul, Turkey — Reyhan Aydoğan
Nagoya, Japan — Rafik Hadfi
May 2016

Contents

Contributors

Bo An School of Computer Engineering, Nanyang Technological University, Singapore, Singapore

Reyhan Aydoğan Interactive Intelligence Group, Delft University of Technology, Delft, The Netherlands; Computer Science Department, Özyeğin University, Istanbul, Turkey

Tim Baarslag Agents, Interaction and Complexity group, University of Southampton, Southampton, UK

Quan Bai Auckland University of Technology, Auckland, New Zealand

Siqi Chen School of Computer and Information Science, Southwest University, Chonqqing, China

Claudia Di Napoli Istituto di Calcolo e Reti ad Alte Prestazioni C.N.R., Naples, Italy

Dario Di Nocera Dipartimento di Matematica e Applicazioni, Università degli Studi di Napoli "Federico II", Naples, Italy

David Festen Interactive Intelligence Group, Delft University of Technology, Delft, The Netherlands

Katsuhide Fujita Institute of Engineering, Tokyo University of Agriculture and Technology, Fuchu, Japan; Faculty of Engineering, Tokyo University of Agriculture and Technology, Tokyo, Japan

Wen Gu Nagoya Institute of Technology, Nagoya, Japan

Jianye Hao School of Software, Tianjin University, Tianjin, China

Masayuki Hayashi Department of Computer Science, Nagoya Institute of Technology, Nagoya, Japan

Koen Hindriks Man Machine Interaction Group, Delft University of Technology, Delft, The Netherlands

Koen V. Hindriks Interactive Intelligence Group, Delft University of Technology, Delft, The Netherlands

Jon Hoffman The University of Tulsa, Tulsa, OK, USA

Enrique de la Hoz University of Alcala, Madrid, Spain

Takayuki Ito Techno-Business Administration (MTBA), Nagoya Institute of Technology, Aichi, Japan

Takayuki Ito Master of Techno-Business Administration, Nagoya Institute of Technology, Nagoya, Japan

Catholijn Jonker Man Machine Interaction Group, Delft University of Technology, Delft, The Netherlands

Catholijn M. Jonker Technical University of Delft, Delft, The Netherlands

Catholijn M. Jonker Interactive Intelligence Group, Delft University of Technology, Delft, The Netherlands

Shinji Kakimoto Institute of Engineering, Tokyo University of Agriculture and Technology, Fuchu, Japan

Shinji Kakimoto Faculty of Engineering, Tokyo University of Agriculture and Technology, Tokyo, Japan

Zahra Khosravimehr Department of Information Technology Engineering, University of Isfahan, Isfahan, Iran

Mark Klein Massachusetts Institute of Technology, Cambridge, MA, USA

Max W.Y. Lam Department of Computer Science and Engineering, The Chinese University of Hong Kong, Shatin, Hong Kong, China

Ho-fung Leung Department of Computer Science and Engineering, The Chinese University of Hong Kong, Shatin, Hong Kong, China

Hoi Tang Leung Department of Information Engineering, The Chinese University of Hong Kong, Sha Tin, Hong Kong, China

Miguel A. Lopez-Carmona Computer Engineering Department, Escuela Politecnica Superior, University of Alcala, Alcala de Henares (madrid), Spain

Xudong Luo Institute of Logic and Cognition, Sun Yat-sen University, Guangdong, China

Ivan Marsa-Maestre University of Alcala, Madrid, Spain

Ivan Marsa-Maestre Computer Engineering Department, Escuela Politecnica Superior, University of Alcala, Alcala de Henares (madrid), Spain

Akiyuki Mori Nagoya Institute of Technology, Aichi, Japan

Shota Morii Nagoya Institute of Technology, Aichi, Japan

Enrique Munoz de Cote Instituto Nacional de Astrofísica, Óptica y Electrónica, Puebla, Mexico

Faria Nassiri-Mofakham Department of Information Technology Engineering, University of Isfahan, Isfahan, Iran

Chi Wing Ng Department of Physics, The Chinese University of Hong Kong, Sha Tin, Hong Kong, China

Tung Doan Nguyen Auckland University of Technology, Auckland, New Zealand

J.B. Peperkamp Delft University of Technology, Delft, The Netherlands

Fenghui Ren School of Computing and Information Technology, University of Wollongong, Wollongong, NSW, Australia

Ansel Y. Rodriguez-Gonzalez Instituto Nacional de Astrofísica, Óptica y Electrónica, Puebla, Mexico

Silvia Rossi Dipartimento di Ingegneria Elettrica e Tecnologie dell'Informazione, Università degli Studi di Napoli "Federico II", Naples, Italy

Swarup Satish Dept. of Electronics and Communication, B.M.S. College of Engineering, Bangalore, India

Sandip Sen The University of Tulsa, Tulsa, OK, USA

Jonathan Serrano Cuevas Instituto Nacional de Astrofísica, Óptica y Electrónica, Puebla, Mexico

V.J. Smit Delft University of Technology, Delft, The Netherlands

Bhargav Sosale School of Computer Engineering, Nanyang Technological University, Singapore, Singapore

Gerhard Weiss Department of Knowledge Engineering, Maastricht University, Maastricht, The Netherlands

Osman Yucel The University of Tulsa, Tulsa, OK, USA

Jieyu Zhan Institute of Logic and Cognition, Sun Yat-sen University, Guangdong, China; School of Computing and Information Technology, University of Wollongong, Wollongong, NSW, Australia

Minjie Zhang School of Computing and Information Technology, University of Wollongong, Wollongong, NSW, Australia

Shuang Zhou Department of Knowledge Engineering, Maastricht University, Maastricht, The Netherlands

Enrique de la Hoz Computer Engineering Department, Escuela Politecnica Superior, University of Alcala, Alcala de Henares (madrid), Spain

Part I
Agent-Based Complex Automated Negotiations

BiTrust: A Comprehensive Trust Management Model for Multi-agent Systems

Tung Doan Nguyen and Quan Bai

Abstract Existing trust and reputation management mechanisms for Multi-agent Systems have been focusing heavily on models that can produce accurate evaluations of the trustworthiness of trustee agents while trustees are passive in the evaluation process. To achieve a comprehensive trust management in complex multi-agent systems with subjective opinions, it is important for trustee agents to have mechanisms to learn, gain and protect their reputation actively. From this motivation, we introduce the BiTrust model, where both truster and trustee agents can reason about each other before making interaction. The experimental results show that the mechanism can overall improve the satisfaction of interaction and the stability of trustees' reputation by filtering out non-beneficial partners.

1 Introduction

Nowadays, Multi-Agent Systems (MASs) have been perceived as a core technology for building diverse, heterogeneous, and distributed complex systems such as pervasive computing and peer-to-peer systems. However, the dynamism of MASs requires agents to have a mechanism to evaluate the trustworthiness other agents before each transaction. Trust was introduced to MASs as the expectation of an agent about the future performance of another agent. Thus, it is also considered as an effective tool to initiate interactions between agents. However, Sen [7] stated that there are not enough research on the establishment, engagement, and use of trust. The establishment of trust can be seen as the "flip side of the evaluation", which focuses on how trustees can gain trust from truster agents actively especially when the subjective opinions are ubiquitous. However, in many real-world circumstances, not only consumers concern about service providers' reputation, but providers also

T.D. Nguyen · Q. Bai (✉)
Auckland University of Technology, Auckland, New Zealand
e-mail: quan.bai@aut.ac.nz

T.D. Nguyen
e-mail: tung.nguyen@aut.ac.nz

© Springer International Publishing AG 2017
K. Fujita et al. (eds.), *Modern Approaches to Agent-based Complex Automated Negotiation*, Studies in Computational Intelligence 674,
DOI 10.1007/978-3-319-51563-2_1

care about who are making requests. For example, online auctions may fail due to the bad behaviours of bidders who refuse to pay after winning.

Obviously, since trust and reputation are crucial in the open and distributed environments, the classic single-sided trust evaluations are inadequate. It implies that truster agents also need to be evaluated for their credibility, i.e., identifying consumers' behaviours. To address this, we introduce the BiTrust (Bijective Trust) model, which enables trustee agents (providers) to reason about truster agents (consumers) to improve the interaction satisfaction with a 2-layer evaluation filter. Specifically, in the first layer, providers evaluate the rating behaviour of the consumers. In the second layer, providers evaluate the utility gain of the transaction respected to their current expectation. The approach not only helps trustee agents to choose a suitable strategy for gaining trust from truster agents but also to protect their reputation actively. This paper steps toward comprehensive trust management in terms of establishing and using trust stated in [7]. The experimental results show several benefits from this model.

The remainder of this paper is structured as follows. Section 2 reviews the most recent updates for trust management. In Sect. 3, we propose and discuss in detail BiTrust model including the trust reasoning of both consumers and providers. An empirical analysis of our method is presented in Sect. 4, it discusses the obtained experimental results. Finally, we conclude the paper in Sect. 5 by summarizing and highlighting the future work.

2 Related Work

In [7], Sen summarised three major aspects in MAS trust management mechanisms, i.e., trust establishment, trust engagement, and trust use. Trust establishment determines the actions and the resources to be invested to establish an agent to be trustworthy to another agent. Trust engagement highlights the intention of rational agents to be active in trust evaluation process and decision making. The use of trust determines how to combine trust and decision to benefit short and long term. Some shortages of existing MAS trust management mechanisms are addressed in [13]. The authors pointed out that the common assumptions of "unlimited process capacity" adopted in MASs may cause the self-damaging reputation problem, no matter how good the trust mechanism is. To address this issue, [12] models reputation as a congestion game and develop a strategy for decision making of trustee agents in resource constraint environment. The paper proposes a trust management model called DRAFT, which can reduce reputation self-damaging problem actively by investigating the capability of trustee itself to decide whether to accept or deny incoming requests. However, it is still weak against requests from truster agents who have biased rating intention after their requests are accepted.

The FIRE [4] model is primarily a trust evaluation model that assumes all agents are honest or unbiased when they report rating and are willing to share the information. These assumptions are not applicable in systems where agents have subjective opinions. The TRAVOS model [9] includes takes lying agents into consideration, but it does not include any strategic reasoning about whether an agent's report is true or false. It has learning ability to filter out inaccurate reputation values, which are better than the pure Bayesian-based method proposed in [10]. However, the opinion filter approaches perform poorly when the percentage of the unfair raters increase. In this situation, without a strategic establishment of trust, trustee agents may fail to interact with potential trusters.

SRAC [1] and DiffTrust [2] performs trust evaluation based on subjective opinions of trusters and advisers. These studies can evaluate trustee based on truster's predefined preferences, but again, trustee agents are independent of truster evaluation process. Some other studies, trusters are also evaluated for their honesty behaviour, but it is from another trusters or the system rather than from trustee agents [6, 9, 11].

Single-sided evaluations could bring the incentives for trustee agents to behave honestly, but not for truster agents, e.g., giving fair ratings. With the BiTrust model introduced in this paper, trustee agents can gain reputation in the protective manner through analysing truster behaviour, estimating the interaction utility, and filtering out non-beneficial transactions.

3 BiTrust: Bijection Trust Management Model

In the BiTrust model, *"trust each other"* is considered as the key to make interaction. It means both the truster and trustee agents need to trust their interacting partners because any damaging reputation will result in the reducing of future interaction. In the BiTrust model, agents have their own preference utility over potential partners. The truster agents' utility function is controlled by parameters related to agent's preferences. The system is assumed to have no centralized database for trust management. Trustee agents evaluate request makers based on their previous behaviours to decide whether to interact with them or not. Thus, BiTrust does not follow prevailing *accept-when-request* assumption, and can be distinguished from the DRAFT approach [12] in that trustee agents accept an interaction based on trust-aware utility gain rather than assessment of their own capability limitation. Below we will give detail definitions used in the model.

3.1 Definitions

For the rest of this paper, we omit the terms trusters and trustees because there is no clear border between them since they both need to be trusted by one another in order to continue the transaction in our model. Instead, the terms "consumer" and "provider" will be used accordingly. Figure 1 shows the conceptual architecture of an agent in the BiTrust model. Each agent (e.g., a_i) has the following five components:

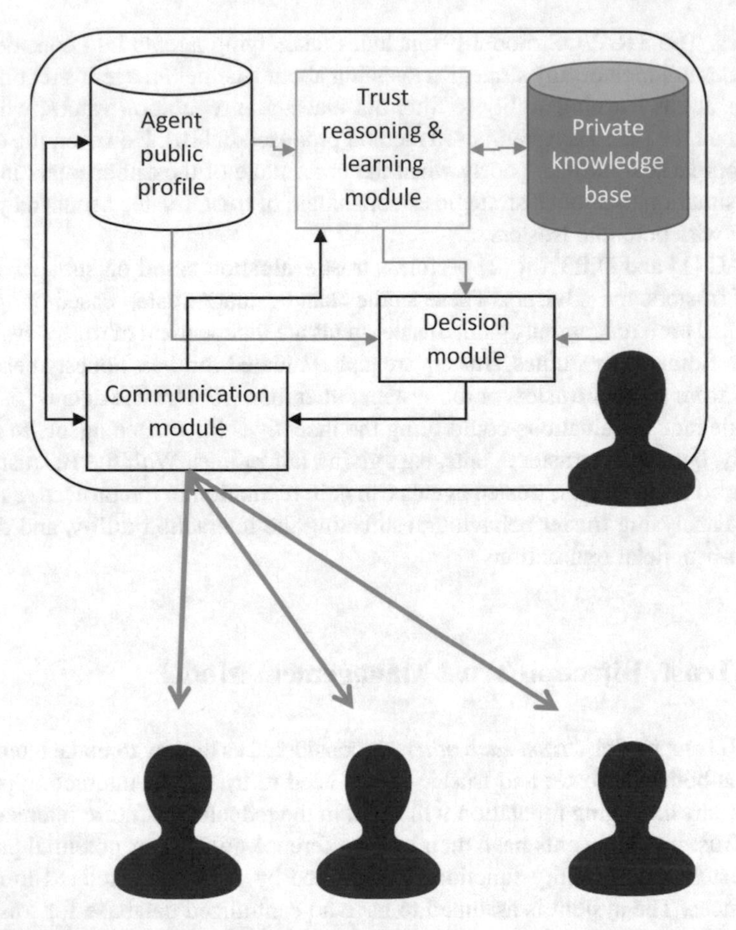

Fig. 1 The architecture of an individual agent in BiTrust

1. *The public profile* is a database contain reputation information (as a provider) and transaction records (H_i) of agent a_i that can be shared with other agents.
2. *The private knowledge base* is a database of each individual agent containing private information or learned experiences.
3. *The trust reasoning and learning module* collects information from databases and communication module to evaluate or learn about the trust of the interacting partner.
4. *The decision module* combines the information from three modules above to decide on which strategies and actions will be applied.
5. *The communication module* helps agent to interact with other agents and surrounding environment.

In this paper, an agent can act as either a consumer or a provider. In addition, an agent can also act as an advisor who gives opinions about other agents. However, we consider an adviser as a type of providers offering reference services which are also rated by consumers. Two advantages of this assumption are: (1) it helps diminish irresponsible advisers; and (2) it brings incentives to advisers to be honest as they are now providers associated with reputation values.

Definition 1 A provider (p_i) is an agent who receives requests from service consumers (See Definition 2), responses and delivers expected service to the consumers.

Providers can be categorized into two main types: reputation sensitive providers (*RSP*) and benefit sensitive providers (*BSP*). The *RSPs* are providers who place their reputation as the highest priority and may sacrifice their benefit in a transaction. For this type of provider, consumer ratings are more important. By contrast, *BSPs* are the ones who consider the benefit as the first priority, they can be considered as ones who are making use of their reputation to maximize their benefit.

Definition 2 A consumer (c_j) is an agent who makes requests and consumes the services provided by providers.

Consumers are categorized into different types, depending on the attributes of service (see Definition 3) they request for. For example, the consumer of delivery service can be assumed to be either a time sensitive, price sensitive or neutral. In other words, consumer types are context-dependent, and there is no public information about consumers' types.

Definition 3 Service reference, $SDes_i$ is the description of service s_i provided by provider p_n. It is a 3-tuple, i.e., $SDes_i = (SType_i, SAtt, SRep)$, contains information about service type (*SType*), service attribute set (*SAtt*) and reputation values ($SRep_i$) of provider p_n.

Service type $SType_i$ identifies the type of a service provided by a provider; *SAtt* is a set of attributes, i.e., $SAtt = (a_1, a_2, \ldots, a_k)$; and *SRep* is a set of reputation value associated to these attributes, i.e., $SRep = (rep_1, rep_2, \ldots, rep_k)$. rep_i is a real value in [0, 1], 0 means worst reputation and 1 means best reputation of attribute a_i.

The reputation information is stored in agent public profile and wrapped into provider's response (see Definition 4). Consumers can query for this information via the communication module. The public reference is partly adapted from CR model [3], which is more suitable to the distributed environments.

Definition 4 Response $Res_{p_jc_i}$ is a message of provider p_j to consumer c_i, consists of service reference (i.e., $SDes_j$) and cost (i.e., $O_{p_jc_i}$), responding to request of consumer c_i. Its format is $Res_{p_jc_i} = (SDes_j, O_{p_jc_i})$.

If provider p_j accepts the request, the offered price $O_{p_jc_i}$ will be a positive real number. Otherwise, $O_{p_jc_i}$ will be negative.

Definition 5 Consumer rating (cr) represents the assessment of performance of a provider on a service, which is a k-tuple, $cr = (r_1, r_2, \ldots, r_k)$, where $r_i (1 \le i \le k)$ indicates the rating of the ith attribute of the service.

As a service contains k attributes, the rating of the consumer also contains k rating values. The rating r_i is a real value ranging from 0 to 1, where 0 and 1 represent the worst and the best performance of ith attribute. For example, a rating can be presented as (0.9, 0.8, 0.75) for attributes (description, quality, delivery). It is advantageous to represent the ratings and reputations with multiple values, because it can help reduce the reputation damage problem mentioned in [12]. For example, when there is a delay in the delivery, only the rating of the delivery attribute will be affected.

Definition 6 Provider feedback on consumer rating is defined as a 2-tuple $fb_{p_jc_i} = (cr, SDes)$, where cr is the rating of consumer (refer to Definition 5) and $SDes$ is the current reputation value of provider (refer to Definition 3).

The feedback from the provider to the consumer will be kept in the consumer's public profile. Those feedback forms the transaction records H_{c_i}. The idea that providers give feedback about the consumers is not new, we can find it in many e-markets, such as eBay. However, the use of providers feedback is unclear in many existing systems. Moreover, the subjectivity of those feedback may double the biased rating problem. For example, if a consumer gives a bad rating for a provider, in turn, the provider can give a consumer a bad rating as revenge. The BiTrust model, however, addresses these issues by not using subjective opinions of providers over consumers' ratings. Instead, rating of a consumer and current status of the provider are combined to construct the feedback about the consumer's rating behaviour and store it in consumers' public database. By doing so, the model leaves evidence for future providers to evaluate the consumer.

3.2 *The BiTrust Protocol*

Figure 2 illustrates the basic procedures to enable BiTrust in which both the consumer and provider need to evaluate each other before the transaction. The providers and consumers must comply the transaction protocol described as the following steps:

1. The consumer c_i evaluates available providers for a specific service, then queries all qualified providers for a specific service. Provider selection is described in Sect. 3.4.

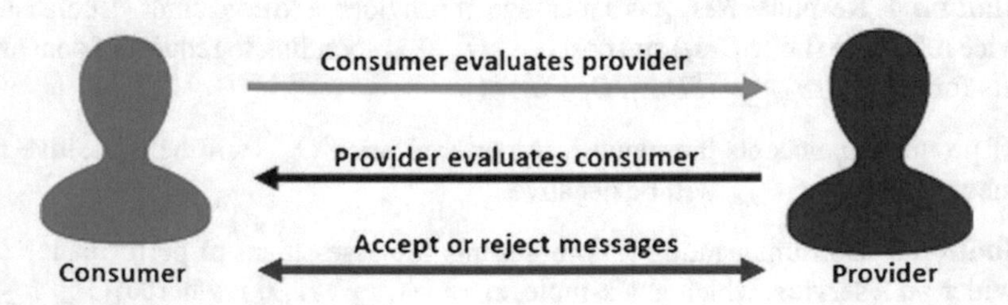

Fig. 2 The BiTrust protocol

2. After receiving a request, the provider p_j evaluates the consumer c_i based on c_i's profile. The provider can then makes a possible offer to consumer c_i or denies the request. The reasoning is described in Sect. 3.3.1.
3. After receiving all offers, the consumer c_i decides whose offer to accept and then carry on the transaction.
4. When the transaction is completed, the consumer c_i evaluates and give the rating of service of provider p_j and p_j stores the rating in the public database.
5. After receiving consumer's rating, the provider p_j generates feedback $fb_{p_j c_i}$ (refer to Definition 6) and give it to consumer c_i to store in the public database.

BiTrust model assumes that providers have a clear understanding of their services. It means providers always know exactly the marginal benefit gained from an offer to a consumer with a specific price and quality. This assumption is important for providers to label the customer types accurately. The mechanism prevents providers from giving subjective rating over consumers thus the model can avoid the doubling biased rating problem. The future providers can investigate the previous feedback to predict whether a consumer is a potential or not.

3.3 *Consumer Behaviour Reasoning*

This subsection will discuss how a provider reasons about a consumer's potential behaviour after receiving a request. The trust reasoning and learning modules (refer to Fig. 1) uses data in the public profile of interacting consumer to reason about the consumer's preference by a behaviour evaluation layer. After that, providers can maximize the utility by choosing suitable offer to satisfy the consumer. Proper offers can help providers get not only the targeted benefit but also trust of consumers. Unlike priority-based trust (PBTrust) [8] where providers do not need to reason about consumers' preference because consumers include attribute weight distribution (or priority) in their requests. However, it is not enough for providers to get the desired ratings from the consumer by solely knowing the consumer's preference because it still depends on rating behaviour of the consumer.

3.3.1 Consumer Behaviour Evaluation Layer

Consumer behaviour in this paper focuses on predicting the ratings of consumers. A provider can discover a consumer's preference through analysing feedback records of the consumer. As introduced in Sect. 3.2, consumers have different weights (priorities) in different attributes of a service. BiTrust model does not treat biased ratings of consumers separately. Instead, it considers the biased ratings as the consequence of the interest weights distribution. With the assumption, the provider reasons about consumers' interest through comparing its rating with the average rating of other consumers on the same provider. If an agent rates the attribute a_m lower than the majority ratings, it has a higher expectation for that attribute. Their interest weight is measured as the ratio between ratings of attribute a_m of consumer c_i with the ratio of the reputation of providers in transaction records of similar service.

$$\omega^{a_m}_{c_i,p_j} = \frac{rep_m}{r_m} \tag{1}$$

Equation 1 indicates the weight ω that a consumer c_i places in an attribute a_m of provider p_j. For example, if the weight for attribute a_1 and a_2 is 0.55 and 0.8 respectively, the consumer is more likely to be a a_1 interested. When $\omega^{a_m}_{c_i,p_j} \geq 1$, the rating of c_i is smaller than the reputation value of attribute a_m or consumer c_i has a higher expectation for this attribute compared to the average. If the consumer has no transaction record, $\omega^{a_m}_{c_i,p_j}$ is set to 1. We perform the analysis over the transaction records of the consumer to get the mean weight $\overline{\omega}^{a_m}_{c_i}$ of consumer c_i for each attribute a_m.

$$\overline{\omega}^{a_m}_{c_i} = \frac{\sum_{p_j \in H_{c_i}} \omega^{a_m}_{c_i,p_j}}{k} \tag{2}$$

In Eq. 2, k is the number of recent transactions of similar service type. $\overline{\omega}^{a_m}_{c_i} \geq 1$ means consumer c_i has positive bias trend to attribute a_m and normally give rating of a_m higher than average, and the provider can benefit from this type of consumer. Likewise, when c_i has a negative bias toward the attribute, provider's reputation of the attribute will take the risk. The value of $\overline{\omega}$ can be normalized to (0, 1) by Eq. 3.

$$\widehat{\omega}^{a_m}_{c_i} = \frac{\overline{\omega}^{a_m}_{c_i}}{\max\{\overline{\omega}^{a_m}_{c_i} | a_m \in SAtt\}} \tag{3}$$

The larger value of $\widehat{\omega}^{a_m}_{c_i}$ indicates the consumer concerns more about the attribute a_m for this service type. By using Eq. 3, providers can obtain some information about consumers' interested attribute but they do not know exactly the weights that a consumer places on their service attributes. However, one important signal is that when the consumer accepts the offer, the current reputations of the service of the providers have passed consumer's trust evaluation. The strategy a_m of provider is to apply to the a_m interested consumer.

The provider's expected rating from a consumer is the subjective expectation of the provider about consumer's future rating to its service based on previous rating behaviour of the consumer to similar service. The expected rating of a consumer to attribute a_i under a_m strategy (service that is delivered for a_m interested consumers) is calculated by using Eq. 4.

$$cr^{a_i}_{e} = \widehat{\omega}^{a_j}_{c_i} \cdot rep_{a_j|a_m}, \tag{4}$$

where $rep_{a_j|a_m}$ is the reputation of attribute a_j calculated through the transaction using strategy for a_m interested consumers. The provider owns the information for calculating $rep_{a_j|a_m}$. The rating prediction should be carried on over all possible strategies of provider, because in real-world situation, providers sometimes cannot guarantee the consumer's satisfaction by giving their services that have the highest rating for attribute a_m.

3.3.2 Provider's Utility Evaluation Layer

After predicting the consumer's rating, the provider can then estimate the utility of the transaction with the consumer. The BiTrust model assumes that providers have a clear understanding of their own services. It means provider knows exactly the marginal benefit gained from the offer. This assumption is important for providers to correctly learn and label the consumer. The BiTrust approach constructs utility function from two components, namely, rating reward and price reward. So provider p_i's utility function is calculated by Eq. 5.

$$U_{p_i}^{c_j} = \alpha \cdot U_{p_i}^{cr} + \beta \cdot \frac{b}{O_{p_j c_i}} \tag{5}$$

In Eq. 5, α and β are predefined weights of rating reward and price reward. They are positive real numbers satisfied $(\alpha + \beta = 1)$. If a provider is reputation sensitive, the value α will be larger than β and likewise. $U_{p_i}^{cr}$ is the rating reward which is calculated by Eq. 6. The rating reward is an estimated rating of the consumer to the current provider. If the estimated rating cr_e of the consumer is higher than pre-defined rating R of the provider, the rating reward is 1 otherwise the reward is -1.

$$U_{p_i}^{cr} = \begin{cases} 1 & \text{if } cr_e \geq \text{R} \\ -1 & \text{otherwise.} \end{cases} \tag{6}$$

In Eq. 6, $cr_e = \{\min\left(rep_m \cdot \omega_{c_i}^{a_m}, 1\right)\}$. The predefined-rating R is the threshold for expected rating of the service provider. The provider uses Eq. 3 to confirm the attribute a_m is interested by the consumer and provide the service with the highest quality of attribute a_m with the belief to maximize possible rating. However, the provider can only confirm accuracy of the strategy after receiving actual consumer rating.

The price reward is calculated by the ratio between benefit b and the offered price $(O_{p_j c_i})$. The benefit b is calculated by the difference between the actual value of the service and the offered price $(O_{p_j c_i})$. The utility function may have negative value if the benefit b is negative.

As mentioned in Sect. 3.2, the learning module will relabel the consumer to training data for better future predictions. In this case, a reinforcement learning approach can be applied as shown in Algorithm 1.

Algorithm 1 Provider learning

Require: predefined utility value U_{p_i}, a set of strategies S_{p_i} containing offer for each consumer.

1: **for** each $RQ_{c_i} \leftarrow$ request of consumer c_i **do**
2: **if** c_i exists in S_{p_i} **then**
3: offer c_i service suggested by S_{p_i}.
4: **else**
5: categorize c_i using Eq. 3
6: calculate $U^{cr}_{pi_i}$ using Eq. 5
7: **if** $U^{cr}_{pi_i} \geq U_{p_i}$ **then**
8: offer service and save c_i to S_{p_i}
9: **else**
10: Reject unaccepted RQ_{c_i}
11: **end if**
12: **end if**
13: $cr_i \leftarrow$ rating from c_i
14: **if** cr_i is as expected **then return**
15: **else**
16: update strategy S_{p_i} for c_i
17: **end if**
18: **end for**

3.4 Service Provider Selection

The consumer preference is important in many contexts involving quality and price because consumers may be indifferent between a provider with a high reputation level and high cost, and a seller with a low reputation level and a low cost if only reputation and price are involved in utility function [10]. Consumer preference also defines a space of acceptance where all belonged providers can be accepted. As shown in Fig. 3, the shaded area is the space of acceptance, the consumer only needs to evaluate utilities of providers in this area.

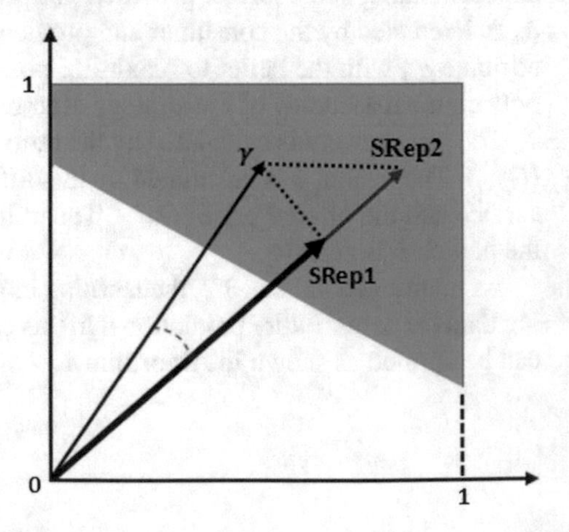

Fig. 3 Trust acceptance space

γ is a consumer defined parameter over the service attributes, which is a vector of attribute weights of a service. O_{c_i} is the expected cost of the predefined attributes. The utility function of the consumer is proportional to provider's reputation suitability (how close the offer O_{p_i} and O_{c_i} are), but inversely proportional to the service price. Equation 7 shows utility function consumer c_i consisting of the suitability indicator with cost.

$$U_{c_i}^{p_j} = \frac{O_{c_i}}{O_{p_j}} \cdot \sum_{i=1}^{k} \gamma_i \cdot rep_i \tag{7}$$

4 Experimental Results and Discussion

To evaluate the effectiveness of the BiTrust model, we have conducted several experiments using simulated data. The hypothesis of the experiment is that the BiTrust model can help providers maintain their reputation better in different circumstances. We assume that there are only two types of provider, i.e., *BSP* and *RSP*, which are managed by two parameters α and β. Providers deliver only one type of service, which consists of three attributes a_1, a_2, and a_3. Each service quality and price can be either: high (H), normal (N), low (L). It means there are 27 different settings for these services. Accordingly, there are three main types of consumers for each attribute a_i interested that influence their rating behaviour with different degrees, namely, high expectation (HE), normal expectation (NE) and low expectation (LE). The higher expectation for an attribute, the lower rating will be given if the consumer is not satisfied. Negative bias consumers are not considered separately. Instead, their expectation are set to high.

The system contains 50 providers and 500 consumers are created with different profiles. Providers and consumers communicate with the protocol defined in Sect. 3.2. To avoid the problem that only some providers are selected by consumers, the experiments prevent providers from accepting requests when they are processing other tasks. The request of consumers can only be done with other providers who are not currently occupied. Each agent assesses its interacting partners separately and individually, no advisor is involved. The obtained results are compared to *accept-when-requested* (AWR) approach to test the hypothesis.

The first experiment tests how profiles influence their success rate of transactions. The system starts and stops when reaches 3000 transactions and repeats 3 times. Table 1 illustrates which consumers is likely to be selected by providers after making requests. Consumers with high expectation for the interested attribute are more likely to be rejected by reputation sensitive providers compared with normal, and low expectation ones. However, they are all likely to be accepted at the similar rate for benefit sensitive providers.

Figure 4a shows the success rate of 2 types of providers, i.e., reputation sensitive (RS) and benefit sensitive (BSPs) providers at the system bootstrap. The result shows that generally, consumers prefer providers who are reputation sensitive. The figures

Table 1 Consumer type acceptance rate

	HE (%)	NE (%)	LE (%)
RS	38.2	68.8	71.4
BS	91.8	95.2	96.6

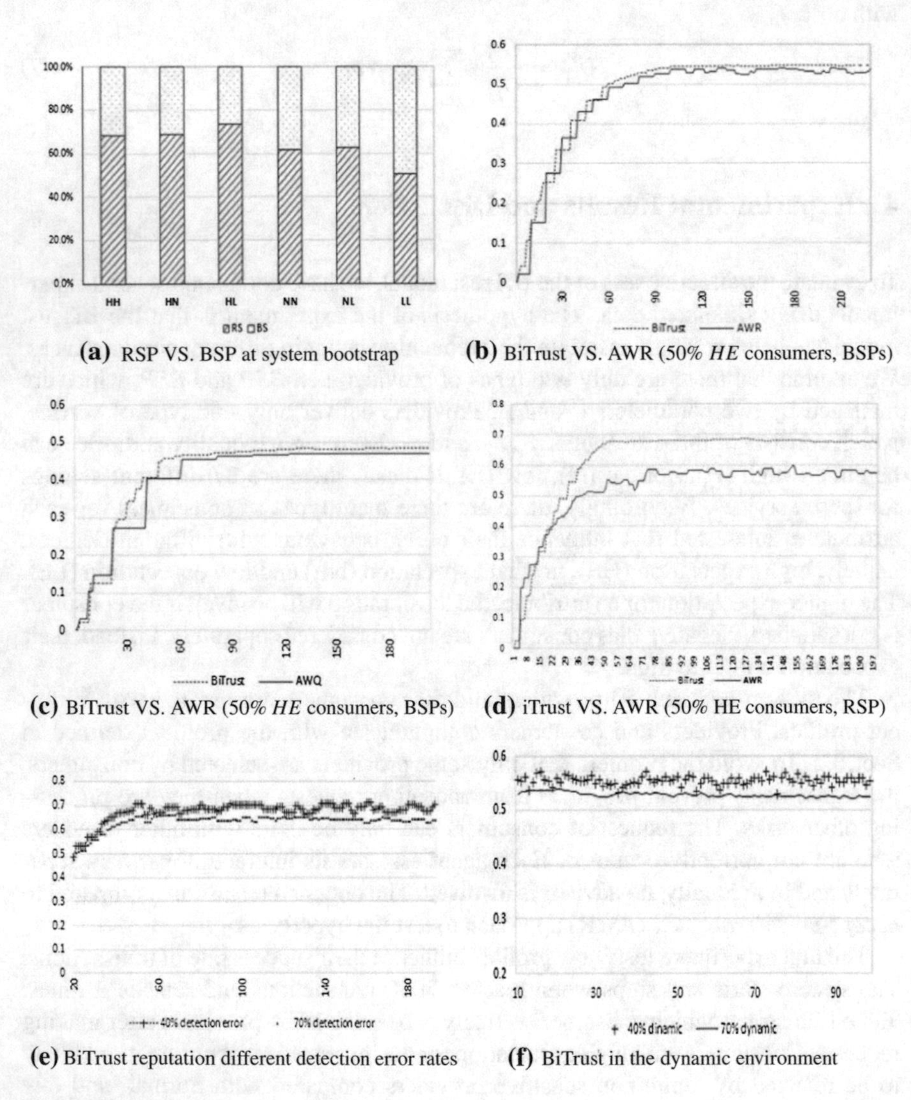

(a) RSP VS. BSP at system bootstrap **(b)** BiTrust VS. AWR (50% *HE* consumers, BSPs)

(c) BiTrust VS. AWR (50% *HE* consumers, BSPs) **(d)** iTrust VS. AWR (50% HE consumers, RSP)

(e) BiTrust reputation different detection error rates **(f)** BiTrust in the dynamic environment

Fig. 4 Experimental results

show that provider with high quality and low price (HL) have the most success rate (73.4%) while there is no significant different when both providers provide low quality, low price (LL) with 50.3 and 49.7%.

The following experiment investigates the reputation values of providers when the number of *HE* consumers increases to 50 and 75% then compare with the *accept-when-requested* model which provider agents do not filter out consumer requests. Figure 4b, c show the experiment result of benefit sensitive providers in environment with 50 and 75% *HE* consumers. The rejection rates of these two cases are: 38.2 and 21.09% with the rejection accuracy is 96.7 and 98.4% respectively. In these settings, providers are benefit sensitive (i.e., $\beta > \alpha$) and that is one reason why the reputation difference between AWR and BiTrust is not very much, however, the marginal reputation gain can be observed in both cases. The rejection rate seems to increase when the number of HE consumers decreases. It can be explained as the providers have better choices when the number of *NE* and *LE* consumers increases. It is obvious that when the number of unfair consumers increases, the reputation drop significantly in these two cases, which is around 0.55 and 0.485 in steady state.

Figure 4d illustrates the reputation in case providers are reputation sensitive. The reputation difference between BiTrust and AWR is larger compared to the case of benefit sensitive providers. In the next experiment, we compare the reputation of providers when there are errors in detecting consumers' interested attribute, i.e., the consumers are interested in different attribute compared to the offered service. Figure 4e shows reputation value in two cases: 40 and 70% detection error *NE* consumers. The reputation values converge to reputation predefined value (threshold) when detection error rate increases. The similar result can be obtained in the case of highly dynamic environment (Fig. 4f) when consumers, who do not have enough transaction records, join and leave system randomly.

Under the AWR setting, the agents' reputation fluctuates significantly. Especially when the number of high expectation consumer increases, the reputation value of those providers are reduced considerably. BiTrust, however, remains stable which can confirm the defined hypothesis. It is also worth to note that, AWR is prone to reputation lag problem [5] more seriously, which is reflected by the scattering plateaus in the figures.

In some circumstances, consumers can be rejected constantly by some providers until those consumers can have better profiles for those providers to accept. It means if a consumer wants to transact with a specific provider, who rejected him earlier, he must carry on the transactions with other providers to update their records.

5 Conclusion and Future Work

In this paper, we have proposed the BiTrust model to enable providers to manage their own reputation through analysing profiles of request makers. With BiTrust, providers can benefit from less fluctuation in their reputation values by denying incoming requests that are not beneficial for the transactions. The model makes use of providers' feedback in consumer profiles, which are likely to be ignored in many existing e-markets. The consumers with bias tendency are more likely rejected in the future interactions unless they consider changing their behaviour to provide more fair ratings. Thus, the model can improve the incentive for providing honest ratings in the system.

The BiTrust model has added a mechanism for providers to select valuable consumers. It is simple to integrate the mechanism to existing trust models which already have single-side trust evaluation. The BiTrust model is one step toward a comprehensive trust management proposed by Sen about how trustee agents can actively gain and protect their reputation. In the future, we will optimize the reasoning algorithm of predicting consumers' behaviour, adjust α and β value dynamically, and investigate deeper the effect on social welfare since the number of transactions may be reduced when both sides concern too much about their interaction partners.

References

1. H. Fang, J. Zhang, M. Şensoy, N.M. Thalmann, Sarc: subjectivity alignment for reputation computation, in *Proceedings of the 11th International Conference on Autonomous Agents and Multiagent Systems*, vol. 3, AAMAS '12 (Richland, SC, 2012), pp. 1365–1366
2. H. Fang, J. Zhang, N.M. Thalmann, A trust model stemmed from the diffusion theory for opinion evaluation, in *Proceedings of the 2013 International Conference on Autonomous Agents and Multi-agent Systems*, AAMAS '13 (Richland, SC, 2013), pp. 805–812
3. T.D. Huynh, N.R. Jennings, N.R. Shadbolt, Certified reputation: how an agent can trust a stranger, in *Proceedings of the Fifth International Joint Conference on Autonomous Agents and Multiagent Systems*, AAMAS '06 (NY, USA, New York, 2006), pp. 1217–1224
4. T.D. Huynh, N.R. Jennings, N.R. Shadbolt, An integrated trust and reputation model for open multi-agent systems. J. Auton. Agents Multi-Agent Syst. **13**(2), 119–154 (2006)
5. R. Kerr, R. Cohen, Modeling trust using transactional, numerical units, in *Proceedings of the 2006 International Conference on Privacy, Security and Trust: Bridge the Gap Between PST Technologies and Business Services*, PST '06 (New York, NY, USA, 2006), pp. 21:1–21:11
6. Z. Noorian, S. Marsh, M. Fleming, Multi-layer cognitive filtering by behavioral modeling, in *Proceedings of the 10th International Conference on Autonomous Agents and Multiagent Systems*, vol. 2, AAMAS '11 (Richland, SC, 2011), pp. 871–878
7. S. Sen, A comprehensive approach to trust management, in *Proceedings of the 2013 International Conference on Autonomous Agents and Multi-agent Systems*, AAMAS '13 (Richland, SC, 2013), pp. 797–800
8. X. Su, M. Zhang, Y. Mu, Q. Bai, A robust trust model for service-oriented systems. J. Comput. Syst. Sci. **79**(5), 596–608 (2013)
9. W. Teacy, J. Patel, N. Jennings, M. Luck, Travos: trust and reputation in the context of inaccurate information sources. J. Auton. Agents Multi-Agent Syst. **12**(2), 183–198 (2006)
10. A. Whitby, A. Jøsang, J. Indulska, Filtering out unfair ratings in bayesian reputation systems, in *Proceedings of the 7th International Workshop on Trust in Agent Societies*, vol. 6 (2004)
11. B. Yu, M.P. Singh, Detecting deception in reputation management, in *Proceedings of the Second International Joint Conference on Autonomous Agents and Multiagent Systems*, AAMAS '03 (NY, USA, New York, 2003), pp. 73–80
12. H. Yu, C. Miao, B. An, C. Leung, V.R. Lesser, A reputation management approach for resource constrained trustee agents, in *Proceedings of the Twenty-Third International Joint Conference on Artificial Intelligence*, IJCAI'13 (AAAI Press, 2013), pp. 418–424
13. H. Yu, Z. Shen, C. Leung, C. Miao, V. Lesser, A survey of multi-agent trust management systems. IEEE Access **1**, 35–50 (2013)

Using Reference Points for Competitive Negotiations in Service Composition

Claudia Di Napoli, Dario Di Nocera and Silvia Rossi

Abstract In a market of services, it is likely that the number of service implementations that exhibit similar functionalities with varying Quality of Service (QoS) will significantly increase. In this context, the provision of a QoS-aware SBA becomes a decision problem on how to select the appropriate services. The approach adopted in the present work is to model both service providers and customers as software agents, and to use automated agent negotiation to dynamically select a set of provider agents whose services QoSs satisfy the customer's requirements. The main features that an automated agent negotiation process should satisfy in order to be applied in service composition are discussed concluding that a multi-issue one-to-many negotiation should be used. In such a setting, we show that using reference points for trading off when different provider agents compete to provide the same service, allows to find (near) Pareto optimal agreements if they exist.

1 Introduction

A *Service-Based Application* (SBA) is a complex business application composed of a number of possibly independent, self-contained, loosely-coupled services, each one performing a specific functionality, and communicating with each other through standard protocols [7]. Such services could be provided by third parties, so the owner

---

C. Di Napoli
Istituto di Calcolo e Reti ad Alte Prestazioni C.N.R., Naples, Italy
e-mail: claudia.dinapoli@cnr.it

D. Di Nocera
Dipartimento di Matematica e Applicazioni, Università degli Studi di Napoli "Federico II", Naples, Italy
e-mail: dario.dinocera@unina.it

S. Rossi (✉)
Dipartimento di Ingegneria Elettrica e Tecnologie dell'Informazione, Università degli Studi di Napoli "Federico II", Naples, Italy
e-mail: silvia.rossi@unina.it

© Springer International Publishing AG 2017

K. Fujita et al. (eds.), *Modern Approaches to Agent-based Complex Automated Negotiation*, Studies in Computational Intelligence 674,
DOI 10.1007/978-3-319-51563-2_2

of the SBA does not control its execution. A service can be characterized also by quality aspects, i.e., by non-functional features referred to as *Quality of Service* (QoS) attributes [22].

In a market of services, customers require SBAs with specific QoS requirements, usually expressed as end-to-end requirements, and several service implementations providing the same functionality are available. So, the selection of services providing the required functionalities with QoS attribute values such that the QoS of the resulting application satisfies the customer's end-to-end QoS requirements is a decision problem. It constitutes an NP-hard problem, complicated and difficult to solve, hence several heuristics approaches have been proposed in the literature.

One of the adopted approaches is to model both service providers and customers as software agents, and to use automated agent negotiation to dynamically select a set of provider agents whose services have QoS values that satisfy the customer's requirements. In an e-commerce based competitive market of services, QoS values are generally bargainable issues, and their adaptive provision can incentivize the selection of a specific service. Moreover, trading off among issue values allows to search for win–win cooperative solutions for the composition in multi-issue negotiation (e.g., paying higher price for a service delivered sooner).

In this work, we discuss the use of software agents and automated negotiation as a means to dynamically select the set of service providers (Sect. 2) competing to provide a service. The main requirements that an automated agent negotiation process should satisfy in order to be applied in service composition are presented (Sect. 3). Such characteristics differ from standard negotiation approaches, so making it difficult to derive optimality properties of the obtained negotiation results. We show that the negotiation process adopted for selecting services for an SBA, has strong similarity with the automated multi-agent multi-issue negotiation solution adopted in [21] to solve a resource allocation problem (Sect. 4). As such, we show that also when more provider agents compete to provide the same service, it is still possible to obtain negotiation outcomes that are (near) Pareto-optimal for the selected set of providers.

2 Composing QoS-Based Services

The service composition process usually starts from an abstract representation of a composition request, we refer to as an *Abstract Workflow* (AW). A simple representation of an AW, also known as the workflow structure, is a directed acyclic graph $AW = (AS, P)$ where $AS = AS_1, \ldots, AS_n$ is a set of nodes, and P is a set of directed arcs. Each node represents an *Abstract Service* (AS_i), i.e., a service description that specifies a required functionality. Each directed arc that connects two nodes represents a *precedence relation* among the corresponding ASs. In order to provide a required SBA, each AS_i has to be bounded to a *concrete service* (we will refer to just as *service*), i.e., a Web service implementing the functionality specified by the corresponding AS_i. Services are provided by different agents, and they may be characterized by quality attributes referring to the service non-functional characteristics.

Typical QoS attributes are: *cost* - the amount that a service requester needs to pay to execute the service; *time* - the execution time between the requests sent and results received; *reliability* - the ability of a service to function correctly and consistently; *availability* - the probability that the service is ready to be invoked; *performance* - related to the service response time and latency; *security* - related to confidentiality and access control aspects. It is becoming of vital importance to take into account the value of these attributes when selecting services to provide an executable workflow, since different customers requiring an SBA may have different expectations and requirements on its end-to-end QoS values. In fact, when requiring SBAs users specify their QoS preferences at the *workflow level* rather than at service level, since they are usually not involved in the service composition process, so they are not aware of how to split a global preference at the level of single services.

2.1 Service Selection

One step of the service composition process is to identify the optimal service selection to meet the user's QoS requirements [22]. In general, service selection can be modeled as a *Multi-dimension Multi-choice Knapsack Problem* (MMKP), which is known to be an NP-hard problem. Exact solutions require long-time computations for large problems, so heuristics approaches are necessary.

By the way, optimization-based approaches consider that the provider's offered values for service QoS attributes are pre-determined and not customizable, but this is unlikely in the context of a dynamic market of services. In fact, the dynamic nature of Web services, and their provision in the Internet-based market of services, require to make the following assumptions:

- the user's QoS requirements may change according to dynamic market demand-supply conditions,
- the set of services available may change in time,
- the QoS values of services may change according to market demand-supply mechanisms, and so they cannot be fixed at the application design-time.

These assumptions make global optimization-based approaches, as the ones proposed in [4] unfeasible in our scenario. For this reason in this work a negotiation-based approach allowing to consider flexible and negotiable QoS attribute values, is adopted. In our approach, it is assumed that service providers are modeled as software agents, we refer to as *Service Providers* (SPs), negotiating with a *Service Compositor* agent (SC) acting on behalf of a user. Negotiation is used for the dynamic selection of the SPs able to provide services whose QoS values, once aggregated, fulfill the user's QoS preferences.

3 Negotiation Requirements for Service Composition

Software agents are a natural way to represent service providers and consumers, and their defining characteristics are essential to realize the full potential of service-oriented systems. Software agents are autonomous problem solving entities, situated in an environment, able to reach their own objectives and to respond to the uncertainty of the environment they operate in, due to flexible decision making capabilities [12]. These characteristics make software agents a useful computational paradigm to model respectively providers that offer services at given conditions, and consumers that require services at other, sometimes conflicting, conditions. Providers and consumers, interacting according to specified protocols and interfaces, have to establish their agreed conditions to respectively provide and consume services. Software agent automated negotiation is one of the approaches adopted for reaching agreements, so it can be used to select services in a service composition. Nevertheless, when negotiation occurs in a realistic market of services, the market characteristics impose specific requirements on the negotiation process, as described in the following subsections.

3.1 One-to-Many

Negotiation usually takes place between two agents willing to come to an agreement on conflicting interests. Most approaches in service composition, that use negotiation mechanisms to select services according to their QoS values, usually apply negotiation for each required service independently from the others relying on bilateral one-to-one negotiation mechanisms [17, 19]. They apply classical negotiation approaches consisting in bilateral interactions of an alternate succession of offers and counteroffers.

In our approach negotiation is used to dynamically select the SPs that offer services with suitable QoS attribute values, but it is assumed that all the agents offering services are involved in the negotiation process. Hence, given an AW composed of n ASs (with $n \geq 2$), and k SPs (with $k \geq 1$) for each of the n ASs in the composition, the number of potential negotiating agents may vary from $n + 1$ to $n * k + 1$ agents, where 1 SC agent is in charge of finding the optimal selection of SPs, according to the QoS user's constraints, to instantiate each AS. Hence, the negotiation is necessarily one-to-many.

3.2 Incomplete Information

In order to prepare an offer $\mathbf{x}_i^t$ at negotiation round t, a service provider agent i uses a set of negotiation strategies to generate values for each negotiated issue. Of course, agents must be equipped with algorithms to evaluate the received and proposed offers.

The value of a specific offer is represented in terms of agent utility. Hence, the utility U_i for an agent i is a function that depends on the specific agent i, and on an offer $\mathbf{x}_j^t$ such as $U_i(\mathbf{x}_j^t) \rightarrow [0, 1]$.

Usually in SBA negotiation the strategies and utility functions adopted by the provider agents are private information. In fact, when SBAs are provided in an open, dynamic and competitive market of services, it is not realistic to assume that their strategies are shared. Furthermore, these strategies may change depending on the market demand-supply trends, so making their shared knowledge unfeasible without causing communication overheads. For these cases, negotiation mechanisms have to be designed so that negotiators can come to an agreement even though they have no prior knowledge (complete or partial) of the utility functions of the other agents involved in the negotiation. Hence, negotiation occurs in an incomplete information setting where agents utility functions, reserve values in terms of utilities, and concession strategies are private information.

The communication occurs only between the SPs and the SC. In addition, also SC constraints on the QoS of the composition may be private. However, even in the case of public constrains, SPs are not able to directly evaluate such constraints since they are not aware of the other offers.

3.3 *Multi-issue*

Negotiation on non-functional parameters of the services composing an SBA is clearly a multi-issue one. In fact, when a service is a component of an SBA, even in the case of a single issue, its value has to be composed with the values of the other services in the composition provided in an independent way, so the negotiation becomes a multi-issue one. More specifically, in the single issue case, the SPs formulate offers containing single issues, but the SC has to evaluate them in an aggregated manner, dealing with a multi-issue evaluation.

In this work, we consider multi-issue SPs offers, hence, an offer made by an SP i at round t is a n-tuple $\mathbf{x}_i^t = (x_{i,1}^t, \ldots, x_{i,m}^t)$, where $x_{i,j}^t$ is a specific value in the domain $\Re$ of the QoS attribute $j \in M$. Multi-issue negotiation is more complex and challenging than single-issue one as the solution space is multi-dimensional, and it is often difficult to reach a Pareto-efficient solution especially in the case of self-interested agents that do not know each other's preferences [14]. Finally, for a single value of utility, different compositions of issues may be available, making the counteroffer process intractable. Typically the strategy of selecting a different configuration of issues values with the same utility value is called *trade-off*.

Typical approaches to multi-issue negotiation are package deal and issue-by-issue [10]. In a package deal negotiation, an offer includes a value for each issue under negotiation, so allowing trade-off to be made among issues. Approaches that adopt issue-by-issue negotiation are based on the assumption that the issues under negotiation are independent. If not, the inter-dependency is addressed by negotiating one issue at a time according to a chosen topology [10]. A general approach to composition should include the case of dependency among issues (e.g., price and time). In this case package deal is the only solution and trade-off is possible among issues. When issues values are interdependent, linear and non-linear utility functions can be used (e.g., Cobb–Douglas, widely used in the economics field [15]).

3.4 Coordinated Interaction

The negotiation mechanism allows to establish a sort of Service Level Agreement (SLA) for QoS-aware SBAs between the SC and the selected SPs. As already said, in a composition of services, also when a single issue is negotiated, its global value is given by the aggregation of the QoS values, each one provided by a service for each AS. That means that the offers received by the SC for a single AS cannot be evaluated independently from the ones received for the other ASs, so a coordination among negotiations for the single abstract services is necessary. A negotiation mechanism for service composition should allow both to negotiate with the SPs providing services for each required functionality in the AW, and also to evaluate the aggregated QoS value of the received offers [5]. So a *coordination step* is necessary. This type of negotiation can be very time-consuming, so the possibility for the SC to concurrently negotiate with the SPs of each AS at the same time is advisable. Generally, a buyer obtains more desirable negotiation outcomes when it negotiates concurrently with all the sellers in competitive situations in which there is information uncertainty and there is a deadline for the negotiation to complete [3]. The coordination step occurs, at the end of each negotiation iteration, when the SC evaluates the aggregation of the received offers in order to allow SPs to adjust their successive offers if an agreement is not reached.

4 The Negotiation Formalization

Let us consider an AW with n ASs (with $n \geq 2$) and m QoS issues (with $m \geq 1$) for each of them, and k SPs (with $k \geq 1$) for each of the n ASs. For each issue j the SC agent has a constraint C_j on the whole AW. The SPs formulate new offers, and the SC evaluates the aggregated value of all considered issues. In this way, it is possible to simulate what happens in a real market of services where a user requesting an SBA does not have information on the SPs strategies. This means that the SC is not able to make single counter-proposals with respect to each received offer, because the change of a value of a particular QoS can impact the constraints to be fulfilled by the QoS of the other services. SC accepts an offer $\mathbf{x}_i^t = (x_{i,1}^t, \ldots, x_{i,m}^t) \in \Re^m$ of the i-th SP if the aggregated value of the offer with the values of the offers for the remaining ASs, satisfies the global constraints, so leading to an agreement.

Definition 1 In case of additive issues, a set of exactly n offers $(\mathbf{x}_1^t, \ldots, \mathbf{x}_n^t)$ is an *agreement* (**A**) at round $t \iff \sum_{i=1}^{n} x_{i,j}^t \leq C_j, \forall j \in m$.

If an agreement is reached with the offers sent at round t, the negotiation ends successfully at that round, otherwise all the offers are rejected and, if $t + 1 < t_{MAX}$, the SC engages all SPs in another negotiation round until the deadline t_{MAX} is reached. Generally, offers are evaluated in terms of agent utility. In a multi-issue negotiation round an agent can either generate a new offer conceding in its utility (i.e., using a concession strategy), or it can select a new offer with the same utility (i.e., using a trading-off strategy in case of dependent issues). In this latter case, these offers belong to the same agent utility curve known as an *indifference curve*.

The i-th SP utility is evaluated in terms of its own offer $\mathbf{x}_i$. In this work we consider evaluation functions that are non-linear. Moreover, the considered evaluation functions are continuous, strictly convex and strictly monotonically increasing in each of the issues.

In general, the utility of an offer $\mathbf{x}_i$ at round t is evaluated as follows:

$$u_i(\mathbf{x}_i, t) = \begin{cases} 0 & \text{if t} = t_{MAX} \text{ and not } (\mathbf{A}) \\ v_i(\mathbf{x}_i) & \text{if } t < t_{MAX} \text{ and } (\mathbf{A}) \end{cases} \quad (1)$$

where, $v_i(\mathbf{x}_i)$ is the evaluation function, **A** is an agreement and t_{MAX} is the deadline.

Here, we explicitly model a collaborative approach among different providers of different services to obtain a win–win opportunity. To enhance the possibility to reach an agreement, each agent may choose the issue values corresponding to a benefit for the other agents on its indifference curve. Indeed, while keeping the same value of utility, the agent chooses to collaborate in order to find an alternative that is better for the others, by trading-off among values. Competition remains among providers of the same service, and it occurs at the concession step.

4.1 The Agents Bidding Strategy

In this work, we focus on the collaborative part of the negotiation, i.e., when agents make trade-off, without considering any concession strategy. In particular, we started from the trading-off strategy proposed in [21] for multi-agent multi-issue negotiation, called the *orthogonal bidding strategy* that was adopted when multiple agents negotiate to distribute units of resources among them. The strategy relies on the possibility of each agent involved in the negotiation to evaluate a so called *reference point* introduced in [20], taking into account the bids of all the other agents involved in the negotiation. Of course, in multi-agent negotiation a reference point cannot be directly computed by applying a one-to-one agent interaction, as in [14]. The same happens in service composition since a single agent offer cannot be used to determine another agent offer because issues are partitioned among more than two agents.

A reference point of an agent, calculated according to the offers of the other agents, as in [21], allows the agent to select, step by step, a new offer on its indifference curve as the point that minimizes the Euclidean distance between the curve and the reference point. Practically, the reference point of an agent represents the desired bid in order to reach an agreement, keeping fixed all the other agents bids. Note that at each step only one agent can send an offer, while the other offers should be kept fixed, so reference points have to be computed one at a time.

In our reference market-based scenario, it is likely that for each AS in the AW more than one SP may issue offers. For this reason, we adopt the heuristic method proposed in [1] to select at each round a set of agents providing a set of promising offers at that round, by assuming that the issues that are negotiated upon are additive (so the workflow structure is not relevant for their composition). The method consists in evaluating the utility of each offer, and in selecting the most promising set of offers, one for each AS, with respect to the global constraints, by considering global constraints as upper bounds for each issue of the composition. So, a promising combination of offers $\mathscr{B} = (\mathbf{b}_1^t, \ldots, \mathbf{b}_n^t)$, one for each AS, is obtained.

Definition 2 A selected offer $\mathbf{b}_k^t$ at round t for the AS_k is the one that maximizes the following equation:

$$\sum_{j=1}^{m} \frac{\underset{\forall x_{i,j}^t \in AS_k}{max}(x_{i,j}^t) - x_{i,j}^t}{\sum_{k=1}^{n} \underset{\forall x_{i,j}^t \in AS_k}{max}(x_{i,j}^t) - \sum_{k=1}^{n} \underset{\forall x_{i,j}^t \in AS_k}{min}(x_{i,j}^t)} \tag{2}$$

where, $max(x_{i,j}^t)$ is the maximum $x_{i,j}^t$ issue value offered by the agent i for the issue j of all the available offers for the AS_k at time t (i.e., $\forall x_{i,j}^t \in AS_k$), while $min(x_{i,j}^t)$ is the corresponding minimum $x_{i,j}^t$ issue value. Equation 2 estimates how good an offer is, by evaluating the QoS values w.r.t. both the ones offered by the other SPs of the same service, by taking as a reference the maximum offered value for that issue, and the QoS values of a possible combination of offers. In fact, the numerator gives an indication of how good the value of each QoS parameter is with respect to the QoS value offered by other SPs of the same AS, and it is then related to the possible aggregated values of the same issue for all the ASs.

Differently from the work of [21], the offers and the SC constraints are private information, so it is not feasible for each SP to compute its own reference point. For these reasons, in our approach, reference points for each AS are calculated by the SC, as a sort of counteroffer, at the coordination step relying on the offers selected for the most promising combination at a given round. In addition, reference points are sent to all SPs providing the same AS, so involving them again in the negotiation even though not selected. So, the SC plays the role of a sort of mediator, since it is the only one that has the necessary information to compute reference points.

A reference point is defined as follows:

Definition 3 The *reference point* for the SPs corresponding to an AS_i and to m issues at round t is:

$$\mathbf{r}_i^t = \left(C_1 - \sum_{k \in N - \{i\}} b_{k,1}^t, \ldots, C_m - \sum_{k \in N - \{i\}} b_{k,m}^t \right) \tag{3}$$

where, $\mathbf{b}_k^t$ is the last bid of agent $k \in N - \{i\}$ selected for the considered combination at that round.

In [21], the authors proved that a set of offers $(\mathbf{x}_1^t, \ldots, \mathbf{x}_n^t)$ is an agreement at round t iff each reference point $\mathbf{r}_i$ for each agent i Pareto dominates the bid of the agent it is calculated for, i.e., $r_{i,j}^t \geq x_{i,j}^t$. Starting from this, the authors proved that, when trading-off among possible offers with the same utility, the orthogonal bidding strategy they propose leads to an agreement that is Pareto optimal and that, if it exists, it is unique. The corresponding theorems were proved for the case of $k = 3$ and $m = 2$. The Definition 3 of reference point is the same as the one defined in [21] with the difference that the constraints C_j (with $j \in M$) are not normalized in the set [0, 1]. So the same theorems apply also in our case provided that reference points are calculated with respect to the set of selected offers at each round, so the Pareto optimality and the uniqueness of the Pareto optimal agreement is referred to the agents providing the set of selected offers $\mathscr{B}$ at the considered round. In fact, different sets of selected offers may lead to different Pareto optimal agreements.

In our approach a reference point, calculated according to Definition 3, is assumed to be the reference point for the entire set of available SPs for each AS at a given round. In this way, all SPs available for each AS are able to negotiate at the successive round by formulating offers based on the value of the reference point, so to avoid discharging offers that may become more promising at successive rounds.

4.2 Weighted Reference Points

When the number of ASs increases, it is undesirable that an SP for a given AS waits for the offers of the others SPs of the remaining ASs to get its reference point, since reference points are computed one at a time. This is even more crucial in an open market of services, since the time spent in negotiation may prevent its use in this scenario. To avoid this, reference points referred to a given round t should be computed relying only on the offers available at the previous round as follows:

Definition 4 The *timed reference point* for the SP_i corresponding to an AS_i at round $t+1$ is:

$$\bar{\mathbf{r}}_i^{t+1} = \left(C_1 - \sum_{k \in N - \{i\}} x_{k,1}^t, \ldots, C_m - \sum_{k \in N - \{i\}} x_{k,m}^t \right) \tag{4}$$

where, for simplicity there is one SP agent for each AS.

Unfortunately, with this definition of reference point, the convergence of the orthogonal bidding strategy is not guaranteed, but it can diverge and lead to an oscillatory behavior. This is due to the fact that reference points are concurrently computed at round t, and used by the SPs to formulate bids at round $t+1$. This prevents the adjustment of bids for each AS, step by step, within the same round that is a prerequisite for the convergence to the agreement. On the other hand, considering the offers at the previous round when computing reference points, is the only way to concurrently negotiate with the SPs for all ASs, so avoiding that the deadlines for each round depend on the number of ASs. To keep the convergence of the bidding strategy, while keeping the possibility to concurrently compute reference points, it is necessary to provide SPs with reference points that allow for different adjustments of bids, in terms of different "weights" that depend on the issue values of the offers with respect to their aggregated values. For this reason, in [6] we introduced a new reference point, named the *weighted reference point* ($\hat{\mathbf{r}}_i^t$) as follows:

Definition 5 The *weighted reference point* for the SP_i corresponding to an AS_i at round $t+1$ is $\hat{\mathbf{r}}_i^{t+1} = (\hat{r}_{i,1}^{t+1}, \ldots, \hat{r}_{i,m}^{t+1})$, with $\hat{r}_{i,j}^{t+1}$ defined as follows:

$$\hat{r}_{i,j}^{t+1} = \frac{x_{i,j}^t}{\sum_{k=1}^n x_{k,j}^t} \cdot \bar{r}_{i,j}^{t+1} = \omega_{i,j}^t \cdot \bar{r}_{i,j}^{t+1} \tag{5}$$

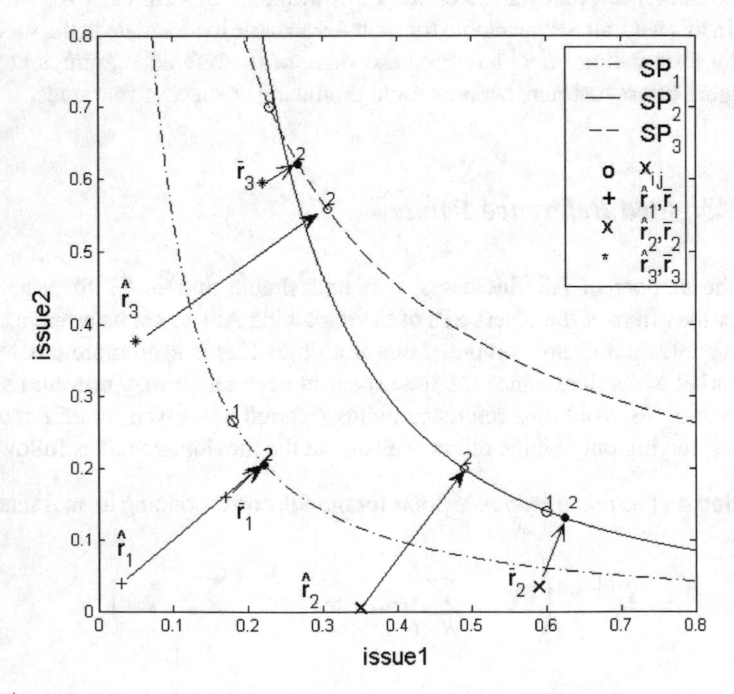

Fig. 1 $\hat{\mathbf{r}}_i^t$ and $\bar{\mathbf{r}}_i^t$ for 2 negotiation rounds

where $\omega_{i,j}^t$ is the weight of the issue value at time t compared to the aggregated value of all the bids for that issue, and $\bar{r}_{i,j}^{t+1}$ is the timed reference point of Definition 4.

In Fig. 1, the behavior of a negotiation in the first two rounds is reported showing reference points and offers in the case of weighted and timed reference points with the same initial configuration. As shown, for SP_1 the $\hat{\mathbf{r}}_1^t$ value corresponds to a scaled version towards the origin of $\bar{\mathbf{r}}_1^t$, since the relative weights of the two issues are comparable in the overall agreement. Instead, for SP_2 and SP_3 the weighted reference points lead to different new bids (number 2) with respect to the case of timed reference points.

According to [6], when trading-off among possible offers with the same utility, the weighted orthogonal bidding strategy leads to an agreement. An investigation of different definitions of weighted reference points for service composition is necessary to verify if Pareto optimality properties can be applied to agreements found when concurrent negotiation is allowed.

5 Simulation Results

Let us consider an AW consisting of 2 ASs and 2 SPs for each AS. Negotiation occurs on two issues (e.g., *issue*1 can be the service execution time, and *issue*2 its cost). SPs utility functions are modeled using the well known Cobb–Douglas functions given by:

$$u_i(\mathbf{x}_i, t) = \gamma (x_{i,1}^t)^\alpha (x_{i,2}^t)^\beta \tag{6}$$

where, α, β and γ are constant factors, with $\alpha \geq 0$, $\beta \geq 0$, and $\gamma > 0$, that are randomly assigned to each agent (and different for each of them), and $x_{i,j} \geq 0$.

In Fig. 2, the evolution of a negotiation execution for the considered experimental setting is shown. In particular, we plotted, for each AS, all SPs issue offers (crosses in the figure) that approach the reference point computed according Definition 3 (empty circles in the figure) for that AS. The best offers selected at each round (filled circles), one for each AS, are used to compute the reference points for the successive round. The negotiation ends successfully with the set of offers respectively sent by SP_1 and SP_3 converging to the Pareto optimal agreement.

In Fig. 3, a different negotiation execution is reported for two provider agents of AS_3, indicating the reference points computed at each round, the corresponding offers respectively sent by the two agents, and the selected offers at each round. As shown, from round 1 to round 4, the offers sent by SP_2 are selected as the most promising ones, while from round 5 to round 10, the offers selected as the most promising ones are those sent by SP_1. The negotiation ends with an agreement including the offer sent by SP_1 at round 10. The possibility to negotiate at each round with all available providers for a given AS, allowed to achieve a Pareto optimal agreement with an agent that would have been discarded since it was not promising at the beginning of the

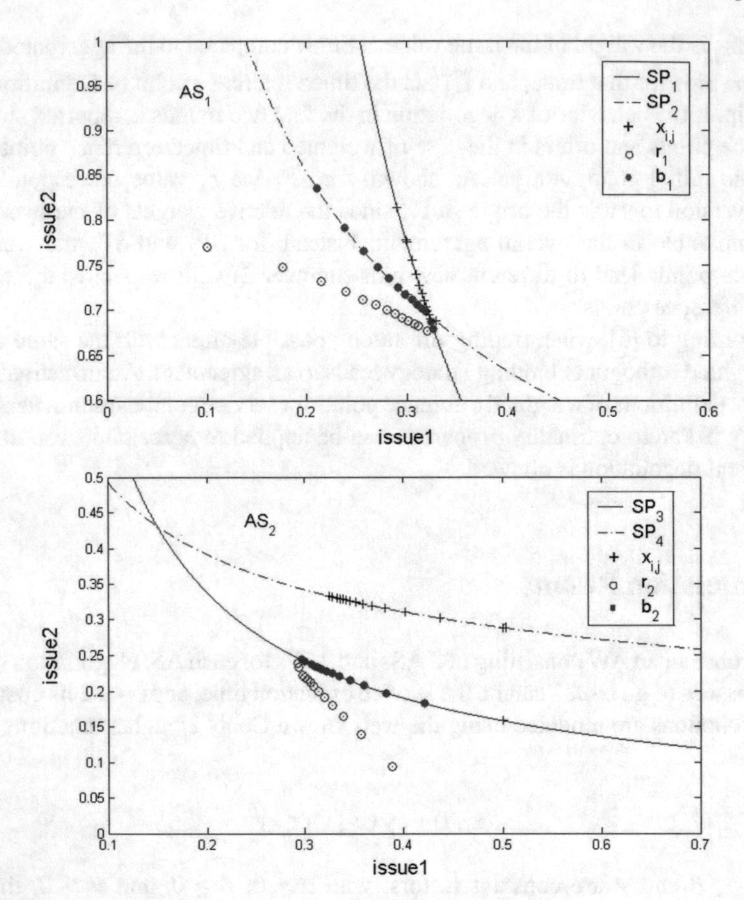

Fig. 2 Negotiation evolution for an AW with 2 ASs, and 4 SPs

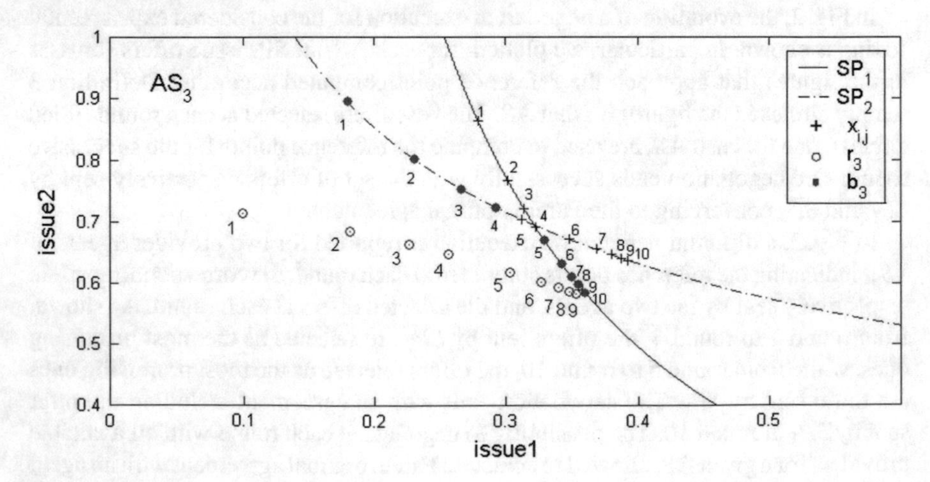

Fig. 3 Offers evolution of the SPs for AS_3

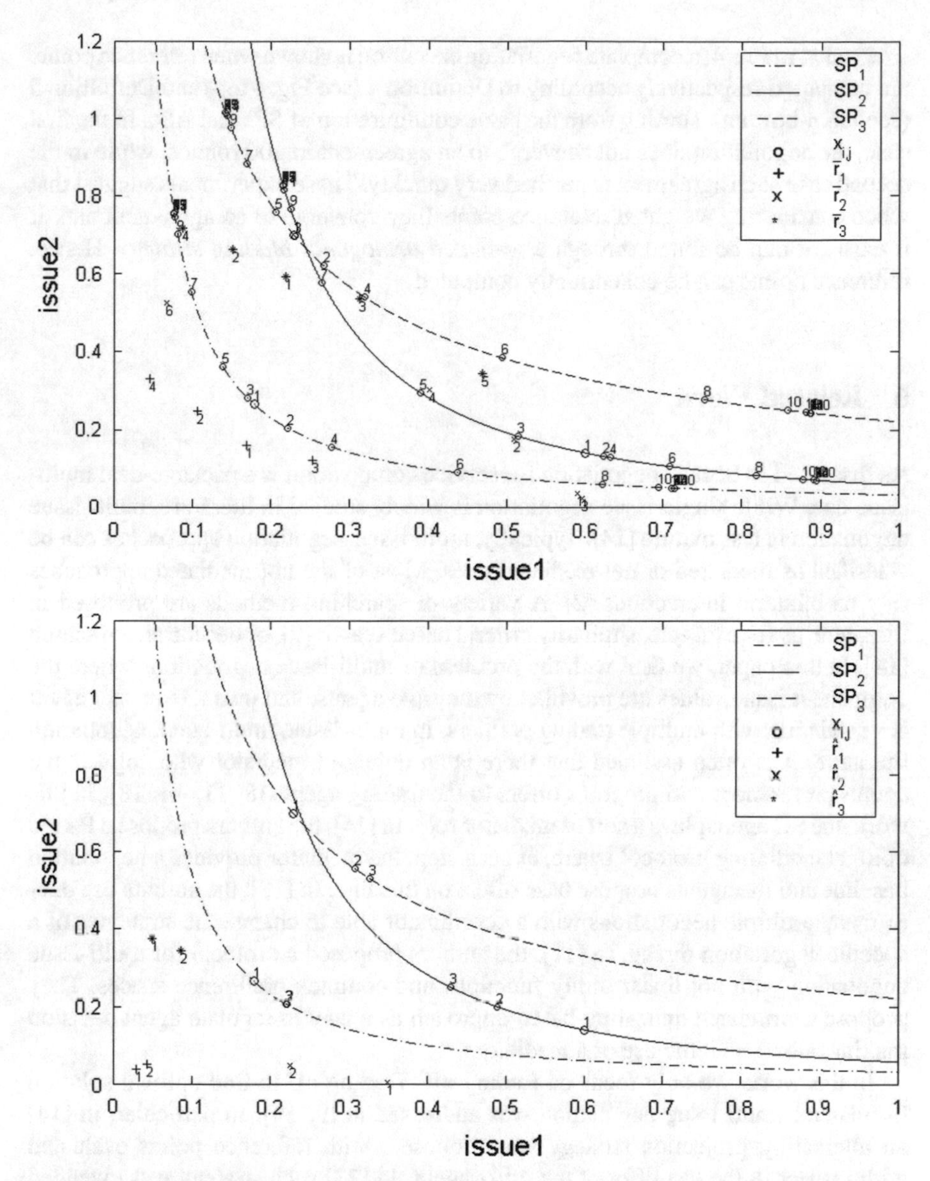

Fig. 4 $\bar{\mathbf{r}}_i^t$ (*top*) and $\hat{\mathbf{r}}_i^t$ (*bottom*) convergence

negotiation. Hence, a reference point computed considering a set of single selected offers at a given round, allows to select a different set of offers at a successive round.

It could happen that an offer for an AS included in a Pareto optimal agreement may be provided by two different SPs, if the indifference curves intersect: in such a case just one of the SPs is randomly selected since the selected agent is not relevant for the Pareto optimality of the agreement.

Finally, in Fig. 4, a complete negotiation execution is shown when reference points are computed respectively according to Definition 4 (see Fig. 4 top) and Definition 5 (see Fig. 4 bottom), starting from the same configuration of SPs and ASs. In the first case, the negotiation does not converge to an agreement in 100 rounds, while in the second case such agreement is reached very quickly. These experiments suggest that when considering weighted reference points they converge to an agreement and, if it exists, it can be found through a *weighted orthogonal bidding strategy*. Hence, reference points can be concurrently computed.

6 Related Work

As discussed in Sect. 3, negotiation for service composition is a package-deal multi-issue one. While single-issue negotiation is widely studied in literature, multi-issue negotiation is less mature [14]. Typically, multi issue-negotiation approaches can be classified as mediated or not mediated ones. Most of the not mediated approaches rely on bilateral interactions [2]. A variety of searching methods are proposed in literature, as for example, similarity criteria based search [9], or decentralized search [14]. In this paper, we deal with the problem of multi-issue negotiations where the component issue values are provided by multiple agents, and thus a requester agent is negotiating with multiple trading partners. In multi-issue, multi-agent negotiation literature, it is often assumed that there is an unbiased mediator who collects the agents preferences and proposes offers to the trading agents [8, 11, 14, 18]. In this work, the SC agent plays a sort of mediator role. In [14], the authors propose a Pareto optimal mediating protocol where, at each step, the mediator provides a negotiation baseline and the agents propose base offers on this line. In [18], the authors use one-to-many multiple negotiations with a coordinator able to change the strategies of a specific negotiation thread. In [11], the authors proposed a protocol for multi-issue negotiation with not linear utility functions and complex preference spaces. They propose a simulated annealing-based approach as a way to regulate agent decision making, along with the use of a mediator.

In this work, we only focus on trading-off. Trading-off to find optimal solution in bilateral multi-issue negotiation was addressed in [9, 14]. In particular, in [14] an alternating projection strategy was proposed, with reference points evaluated with respect to the last offer of the other agent. In [23] such strategy was extended to the multi-agent case, by evaluating reference points as a mean sum of all the offers at each step. Differently from our case, in [23] an agreement corresponds to a single point in the negotiation space, and weights are the same for all the agents. In [9], the authors used the notion of fuzzy similarity to approximate the preference weights of the negotiation opponent in order to select the most similar offer to the last received offer in a pool of randomly generated trade-off offers. In [8], the authors present a constraint proposal method to generate Pareto-frontier of a multi-issue negotiation corresponding to a given reference point. In practice, the mediator adjusts a hyperplane according to predetermined reference points, until the agents most preferred alternatives on the hyperplane coincide. By choosing reference points on the line connecting the agent global optima, Pareto optimal points are produced, and the mediator's problem has a solution when the number of issues is either two or any odd number greater than two [13]. In [21], the authors present an automated multi-agent multi-issue negotiation for allocation of unit resources, similar to our case. The proposed bidding strategy requires that at each round the agents make bids in a sequential order in order to compute a reference point for each agent involved in the negotiation. In our approach, reference points are calculated for each set of provider agents providing a specific functionality required in a service composition.

Generally, a buyer gets more desirable negotiation outcomes when it negotiates concurrently with all the sellers in competitive situations in which there are information uncertainty and deadlines [16]. A model of concurrent negotiation was addressed in [2], where agents are allowed to make counter-proposal without having received proposals from all other trading partners. In [16], the multiple negotiation threads still happen in the same negotiation round, as in our case, but the heuristic methods used by the negotiation coordinator strongly depend on history information about trading partners and negotiation environment. In our dynamic market based scenario, past information is not always relevant to drive negotiation.

7 Conclusions

In this work, we discussed the main features that make software agent negotiation a suitable approach to select services depending on their QoS attribute values. As described, when service provision occurs in a competitive market of service providers, the adopted negotiation model has to meet specific requirements to be applied in a service composition problem. Since negotiation occurs among a user requesting an SBA and the providers available to deliver the appropriate service components, usually characterized by multiple QoS attributes, negotiation is a multi-agent and multi-issue one. For this type of negotiation it is more difficult to derive theoretical understanding of its behavior, and more crucially to define when agreements that are Pareto optimal can be found. In this work, we refer to a scenario where a composition of services have to be delivered with QoS value satisfying a user's request, assuming that for each component service more providers are available on the market, and they may provide the same service with different QoS additive values. In this scenario, we proposed a variation of the orthogonal bidding strategy based on the approach presented in [21], and showed how it allows to find an agreement, if it exists, that is Pareto optimal. Furthermore, the possibility to negotiate at each round with all available providers for each abstract service in the composition, allows to achieve a Pareto optimal agreement with a provider agent that would have been discarded according to the adopted heuristics, since it was not promising at the beginning of the negotiation. Hence, a reference point computed considering a set of selected offers at a given round, allows to select a different set of offers at a successive round.

In addition, by introducing a weighted reference point, we show that it is still possible to find an agreement also in the case the Service Compositor concurrently computes all the reference points for each Abstract Service. This allows to avoid making the length of negotiation depending on the number of the Abstract Services composing the Abstract Workflow, that is the case when computing reference points one at a time. This aspect is important when adopting negotiation for service composition. This is even more crucial when the considered reference scenario for service composition is an open market of services, since the time spent in negotiation may prevent its use in these settings.

References

1. M. Alrifai, T. Risse, Combining global optimization with local selection for efficient qos-aware service composition, in *Proceedings of the 18th International Conference on World Wide Web* (ACM, 2009), pp. 881–890
2. B. An, K.M. Sim, L.G. Tang, S.Q. Li, D.J. Cheng, Continuous-time negotiation mechanism for software agents. IEEE Trans. Syst. Man Cybern. Part B: Cybern. **36**(6), 1261–1272 (2006)
3. B. An, V. Lesser, K. Sim, Strategic agents for multi-resource negotiation. Auton. Agents Multi-Agent Syst. **23**(1), 114–153 (2011)
4. D. Ardagna, B. Pernici, Adaptive service composition in flexible processes. IEEE Trans. on Softw. Eng. **33**(6), 369–384 (2007)
5. C. Di Napoli, D. Di Nocera, P. Pisa, S. Rossi, A market-based coordinated negotiation for qos-aware service selection, in *Designing Trading Strategies and Mechanisms for Electronic Markets*, LNBIP, vol. 187 (Springer, 2014), pp. 26–40
6. C. Di Napoli, D. Di Nocera, S. Rossi, Computing pareto optimal agreements in multi-issue negotiation for service composition, in *Proceedings of the 2015 International Conference on Autonomous Agents and Multiagent Systems*, AAMAS '15 (2015), pp. 1779–1780
7. S. Dustdar, W. Schreiner, A survey on web services composition. Int. J. Web Grid Serv. **1**(1), 1–30 (2005)
8. H. Ehtamo, R.P. Hamalainen, P. Heiskanen, J. Teich, M. Verkama, S. Zionts, Generating pareto solutions in a two-party setting: constraint proposal methods. Manag. Sci. **45**(12), 1697–1709 (1999)
9. P. Faratin, C. Sierra, N. Jennings, Using similarity criteria to make issue trade-offs in automated negotiations. Artif. Intell. **142**(2), 205–237 (2002)
10. S.S. Fatima, M. Wooldridge, N.R. Jennings, On optimal agendas for package deal negotiation. AAMAS **3**, 1083–1084 (2011)
11. T. Ito, H. Hattori, M. Klein, Multi-issue negotiation protocol for agents: exploring nonlinear utility spaces, in ed. By M.M. Veloso. IJCAI (2007), pp. 1347–1352
12. N.R. Jennings, An agent-based approach for building complex software systems. Commun. ACM **44**(4), 35–41 (2001)
13. M. Kitti, H. Ehtamo, Analysis of the constraint proposal method for two-party negotiations. Eur. J. Oper. Res. **181**(2), 817–827 (2007)
14. G. Lai, K. Sycara, A generic framework for automated multi-attribute negotiation. Group Decis. Negot. **18**(2), 169–187 (2009)
15. A. Mas-Colell, M.D. Whinston, J.R. Green, Microeconomic Theory (Oxford University Press, Oxford, 1995)
16. T.D. Nguyen, N.R. Jennings, A heuristic model of concurrent bi-lateral negotiations in incomplete information settings, in *IJCAI* (2003), pp. 1467–1469
17. S. Paurobally, V. Tamma, M. Wooldrdige, A framework for web service negotiation. ACM Trans. Auton. Adapt. Syst. **2**(4) (2007)

18. I. Rahwan, R. Kowalczyk, H.H. Pham, Intelligent agents for automated one-to-many e-commerce negotiation (IEEE Computer Society Press, 2002), pp. 197–204
19. F. Siala, K. Ghedira, A multi-agent selection of web service providers driven by composite qos, in *Proceedings of 2011 IEEE Symposium on Computers and Communications (ISCC)*. IEEE (2011), pp. 55–60
20. A. Wierzbicki, The use of reference objectives in multiobjective optimization, in *Multiple Criteria Decision Making Theory and Application*, LNEMS, vol. 177 (Springer, 1980), pp. 468–486
21. M. Wu, M. de Weerdt, H. La Poutré, Efficient methods for multi-agent multi-issue negotiation: allocating resources, in ed By J.J. Yang, M. Yokoo, T. Ito, Z. Jin, P. Scerri. *Principles of Practice in Multi-Agent Systems*, LNCS, vol. 5925 (Springer, 2009), pp. 97–112
22. T. Yu, K. Lin, Service selection algorithms for web services with end-to-end qos constraints, in *Proceedings of the IEEE International Conference on E-Commerce Technology*. IEEE (2004), pp. 129–136
23. R. Zheng, N. Chakraborty, T. Dai, K. Sycara, Multiagent negotiation on multiple issues with incomplete information: extended abstract, in *Proceedings of AAMAS* (2013), pp. 1279–1280

A Cooperative Framework for Mediated Group Decision Making

Miguel A. Lopez-Carmona, Ivan Marsa-Maestre and Enrique de la Hoz

Abstract In this work we consider a group decision making problem where a number of agents try to reach an agreement through a mediated automated negotiation process. Each participating agent provides her preferences over the sets of contracts proposed by the mediator in successive mediation steps. Then, individual preferences are aggregated to obtain a group preference function for choosing the most preferred contract. The negotiation process involves a set of mediation rules to explore efficiently the alternatives space, which is derived from the Generalized Pattern Search non-linear optimization algorithm. A particularly notable feature of our approach is the inclusion of mechanisms rewarding the agents for being open to alternatives other than simply their most preferred. The proposed negotiation framework avoids selfish behavior and improves social welfare. We show empirically that our approach obtains satisfactory results under smooth non-linear utility spaces.

1 Introduction

Group decision making is among the most important decision processes in technical, social and economic contexts, and cooperation is at the heart of fair and optimal group decisions. A typical framework for group decision making support consists on a mediator, a set of alternatives, a set of preference functions stating the support that an agent gives to each alternative, and a set of payoffs stating the reward that an agent will get if certain alternative is finally selected [16]. The role of the mediator is to privately aggregate these preference functions and select the most supported

M.A. Lopez-Carmona (✉) · I. Marsa-Maestre · E. de la Hoz
Computer Engineering Department, Escuela Politecnica Superior,
University of Alcala, Campus Politecnico, 28871 Alcala de Henares (madrid), Spain
e-mail: miguelangel.lopez@uah.es

I. Marsa-Maestre
e-mail: ivan.marsa@uah.es

E. de la Hoz
e-mail: enrique.delahoz@uah.es

© Springer International Publishing AG 2017

K. Fujita et al. (eds.), *Modern Approaches to Agent-based Complex Automated Negotiation*, Studies in Computational Intelligence 674,
DOI 10.1007/978-3-319-51563-2_3

alternative. The strategies used by the agents when revealing their preferences will strongly influence on the final agreement [17].

Existing research recognizes the critical role played by aggregation functions to enforce cooperation in mediated group decision making [1, 20]. In order to avoid strategic manipulation, Yager [19] develops a one-shot automated negotiation mechanism that privileges agents who are not totally self-interested by using a weighted aggregation operator. Several studies have reported mechanisms to define complex forms of social welfare by means of Ordered Weighted Average (OWA) operators [2–4]. For instance, in [19] a set of mediation rules are defined which allow for a linguistic description of social welfare using fuzzy logic.

A number of researchers have built iterative mechanisms to reach agreements among multiple participants [7–9, 13, 14, 18]. They define bargaining mechanisms to obtain agreements by using fair direction improvements in the joint exploration of the negotiation space [5, 6, 12]. The mediator proposes a set of alternatives and agents provide their utility gradients. Then, the mediator proposes a new set of alternatives in the bisector or in an arbitrary direction which is considered fair enough. Unfortunately, these mechanisms are prone to untruthful revelation to bias the direction generated by the mediator.

To avoid strategic manipulation, in [10] agents are restricted to express their preferences using a minimum number of positive votes for the different alternatives. Then, the mediator looks for Pareto improvements using a mediated negotiation protocol based on a distributed non-linear optimization mechanism.

To the best of our knowledge there are no results in the literature regarding how more complex aggregation operators impacts strategic manipulation, social welfare and optimality in mediated group decision making.

The aim of our paper is to propose a multi-party negotiation framework that rewards those agents who are not totally self-interested. It relies on a distributed exploration protocol based on the *Generalized Pattern Search* optimization algorithm (GPS) [11], and a set of mediation rules that rewards truthful revelation of preferences. Our proposal considers the iterative improvement of potential solutions to the problem.

The structure of the paper is as follows. In Sect. 2 we describe the problem. Then, in Sect. 3 we present the adapted GPS algorithm to look for Pareto improvements. We present the negotiation protocol, the aggregation operator and the alternatives search process. In Sect. 4 we present the experiments and the results obtained. Finally conclusions and further work are presented in Sect. 5.

2 Problem Description

We consider an iterative mediation protocol performed concurrently by a set of agents $a_1, a_2, \ldots, a_n$ and a mediator. The aim of the agents is to reach an agreement on a contract which is defined by a set of issues in the real domain. The mediator proposes a set of contracts, and the agents express privately their preferences. We assume non-linear preferences which may take values from the real or integer domain. Then, the mediator aggregates the preferences for each contract and selects a winner. Based on the winner contract a new set of contracts is proposed to the agents in order to improve the search process. This iterative process ends with a globally accepted contract.

2.1 Negotiation Domain and Agents' Preferences

Each contract $c \in C$ is a vector defined by a finite set of issues $\{x_i | i = 1, \ldots, k\}$ in the real domain. Thus, the negotiation domain can be denoted $C \in \mathbb{R}^k$. We denote by $V_{a_p}(c)$ the payoff obtained by agent a_p if the output of the negotiation process is c. Additionally, $A_{a_p}(c)$ defines the degree to which a_p supports c. Without loss of generality, we consider $A_{a_p}(c) \in [0, 1] \forall a_p, c$, being 1 the highest support. Note that $V()$ is a private function stating the reward achieved by the agent, while $A()$ reflects the information revealed to the mediator.

To model non-linear preferences we use *Bell functions* [15]. Bell functions capture the intuition that agents' preference for a contract usually decline gradually with distance from their ideal contract. In addition, they provide the capability of configuring different negotiation scenarios with different complexity degrees.

A *Bell function* is defined by a center c, height h, and a radius r. Let $\| s - c \|$ be the euclidean distance from the center c to a contract s, a *Bell function* is defined as

$$fbell(s, c, h, r) = \begin{cases} h - 2h\frac{\|s-c\|^2}{r^2} & if \ \| s - c \| < r/2, \\ \frac{2h}{r^2}(\| s - c \| - r)^2 & if \, r > \| s - c \| \geq r/2, \\ 0 & \| s - c \| \geq r \end{cases}$$

where the *reward function* is an aggregation of Bell functions

$$V_{a_p}(s) = \sum_i^{nb} fbell(s, c_i, h_i, r_i) \, .$$

The variable nb represents the number of bells used to define each reward function. The complexity of the reward functions can be modulated by varying c_i, h_i, r_i and nb.

Figure 1 shows a proof of concept scenario with reward functions in a bidimensional contract space $[0, 100]^2$. The Social welfare plot presents the sum of the agents' rewards, and the Agents 3 and 4 plot the sum of Agents' 3 and 4 rewards. Each reward function has a global optimum and two local optima in different quadrants of the contract space. The maximum social welfare is around 1.8 for the contract [50, 50], but the highest reward for an agent is close to one of the corners in the contract space

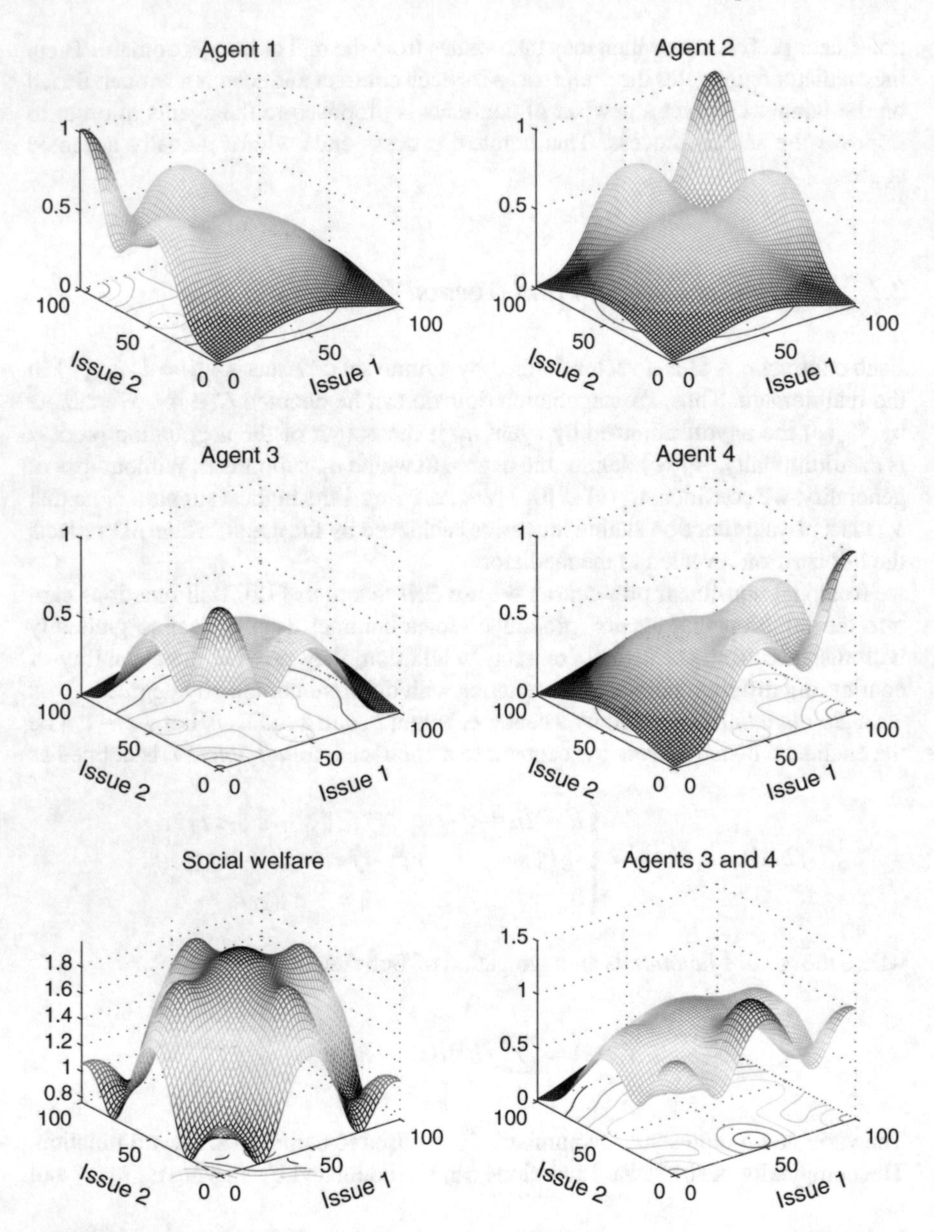

Fig. 1 Reward functions for a *proof of concept negotiation scenario*

(i.e. [0, 100], [100, 100], [0, 0], [100, 0]). We used this proof of concept scenario in our tests in order to demonstrate the general properties of our approach.

2.2 Negotiation Protocol

We propose a distributed adaptation of the GPS optimization algorithm, which serves to perform a distributed search through the contract space. For simplicity and without losing generality, we will assume a negotiation domain with contracts formed by two issues (i.e. $C \in \mathbb{R}^2$) in the description of the protocol. The extension to more issues is straightforward.

At any given round t, the mediator builds a set of five contracts $C^t = \{c_0^t, c_0^t + \Delta^t \times e_j \mid j = \{1, \ldots, 4)\}$, where e_j is the basis vector and $\Delta^t \in \mathbb{R}$ is a step-length parameter. Thus, $C^t = \{c_0^t, c_1^t, c_2^t, c_3^t, c_4^t\}$ is formed by a contract c_0^t and a set of four contracts which are centered around c_0^t. For two issues, a commonly used basis is: $e_1 = \{1, 0\}$, $e_2 = \{0, 1\}$, $e_3 = \{-1, 0\}$ and $e_4 = \{0, -1\}$. For a number of issues n we would need at least a basis of length $2n$.

In the first round, the mediator selects a random contract c_0^0 and builds the other four contracts using an initial Δ^0. This set of contracts C^0 is sent to the agents and then negotiation starts:

1. Each agent a_p provides the mediator a set of supports $A_{a_p}(C^t)$ for the different contracts in C^t (C^0 in the first round). We recall that the support information is private to the agents and only known to the mediator, and that $A_{a_p}(C^t)$ may not reflect the real reward $V_{a_p}(C^t)$ achieved by the agent.
2. The individual preferences for each contract c_i^t are aggregated by the mediator using the aggregation operator D

$$D_i^t(A_1(c_i^t), \ldots, A_n(c_i^t))$$

to obtain the set of group preferences $Gp^t = \{D_i^t | i = 0 \ldots 4\}$. We shall refer to this as the **aggregation of preferences** step.
3. The mediator selects the most supported contract depending on the values in Gp^t. For instance, it can be selected the most supported contract, or we could consider the uncertainty of the selection process, and use a probabilistic function to select the winner contract. We shall refer to this as the **contract selection** step. Negotiation ends at this stage if a maximum number of rounds has been reached. Otherwise, the negotiation protocol goes to step 4.
4. If the selected contract in step 3 is c_0^t, then Δ^t is reduced by half; otherwise, the step-length parameter is increased by two. In both cases, the selected contract will be the new central contract c_o^{t+1} in the next round. If Δ^{t+1} is deemed sufficiently small then negotiation ends with the winner contract as the final agreement. Otherwise, the process goes to step 5.

5. Based on the group preferred contract c_0^{t+1} and Δ^{t+1}, the mediator obtains a new set of contracts C^{t+1} which is sent to the agents, and then the process goes to step 1 again.

We assume that the negotiation process always ends with an agreement. Negotiation ends when Δ^t is below a predefined threshold, either when group preference is above a threshold or when the number of rounds expires.

At each stage an agent provides her support for the different contracts, which comes determined by her underlying payoff function and any information available about the previous stages of the negotiation. The process of choosing the specific support for the different alternatives at each round of the negotiation then constitutes a participating agent's strategy. An important consideration in an agent's determination of their strategy are the rules and procedures used in the negotiation process.

Figure 2 shows an example of negotiation process with 2 agents and 35 rounds. The first plot illustrates the movement, expansion and contraction of the set of contracts. Each point is a contract such that its diameter is proportional to the step-length used by the mediator. We can see how the points get smaller as we get closer to the final agreement. In the example, the maximum social welfare is around the central point. The second plot shows the evolution of the agents' rewards. The third plot presents the evolution of the group preferences. Each column comprises five points representing the group preferences for each contract in C^t.

In the following we shall describe the implementation of the negotiation protocol steps outlined above.

2.3 Aggregation Operator for Adapted GPS (GPSao)

Our aim was to define an aggregation mechanism such that agents were prone to truthfully reveal their preferences and mitigate strategic manipulation. We needed an operator that considered the score distribution within the set of contracts in C^t. We called this operator the *GPS aggregation operator* or **GPSao**, which is a based on a weighted sum of the individual preferences:

$$D_i^t = w_{a_1} A_{a_1}(c_i) + w_{a_2} A_{a_2}(c_i) + \ldots + w_{a_n} A_{a_n}(c_i)$$

where the weights for each agent a_p are computed as follows:

$$m_p = \max\{A_{a_p}(c_0^t), A_{a_p}(c_1^t), ..., A_{a_p}(c_4^t)\}, \tag{1}$$

$$S_p = 1 - \frac{\sum_{i=0}^{4}(m_p - A_{a_p}(c_i^t))}{4m_p}, \tag{2}$$

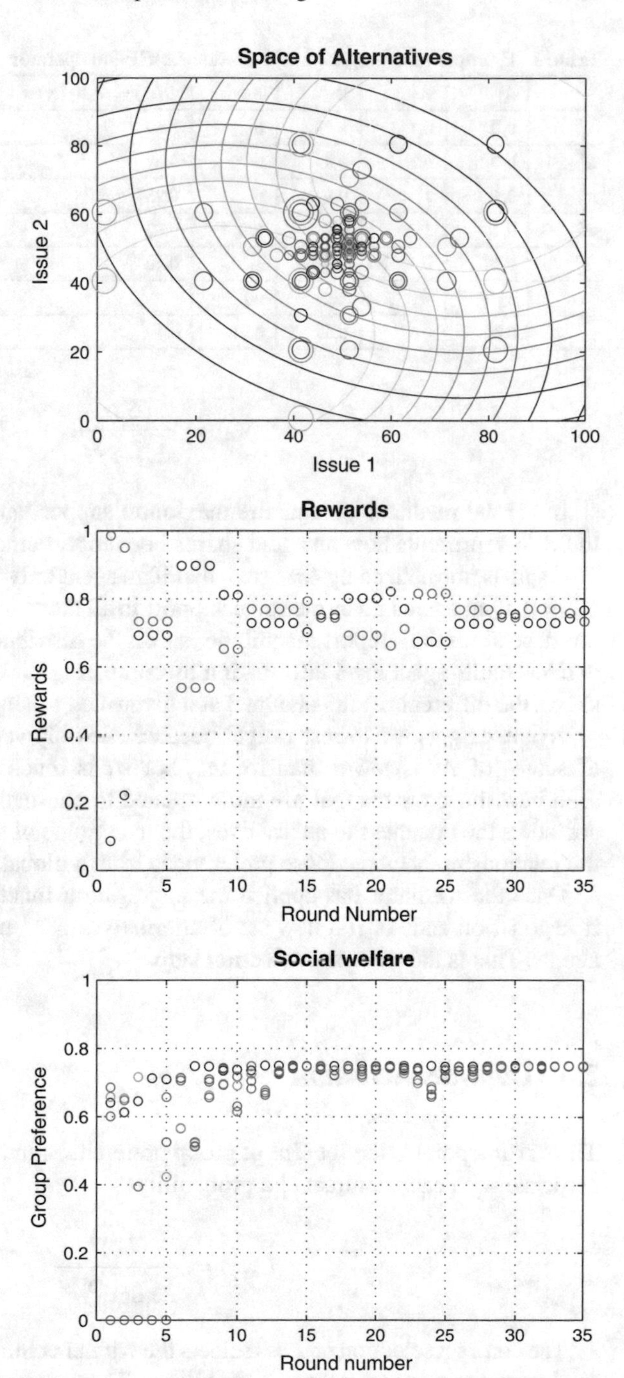

Fig. 2 Example of negotiation process: expansion and contraction of alternatives, evolution of rewards and evolution of social welfare

Table 1 Example of aggregation of preferences: GPSao operator

	A_1	A_2	A_3	A_4	$D(c_i^t) = w_1A_1(c_i) + w_2A_2(c_i) + \ldots + w_nA_n(c_i)$
x_0^t	0, 2	0.7	1	0	0.69
x_1^t	0, 3	0.7	0.8	0	0.64
x_2^t	0, 2	1	0.8	0	0.73*
x_3^t	0, 1	1	0.7	0, 1	0.67
x_4^t	0, 01	0, 4	0.6	1	0.40
	w_1	w_2	w_3	w_4	
	0.23	0.37	0.39	0.01	

$$w_{a_p} = \frac{S_p}{\sum_{j=1}^{n} S_j}. \tag{3}$$

In (1) the mediator obtains the maximum support an agent gives to a contract. In 38 S_p represents how an agent shares her support among the different contracts. This sum is normalized by $4m_p$ such that if an agent only votes for one contract, then $S_p = 0$. If an agents gives the same support to all the contracts then $S_p = 1$. S_p does not depend on the support magnitude but on the distribution of scores. Finally, w_{a_p} extracts multi-agent level information by comparing the different S_p values. Table 1 shows the different results obtained at a given step t using GPSao.

Agent A_4 gets the lowest weight because she only votes for x_3 and x_4. The sum of scores of A_1 is lower than for A_4, but w_1 is much higher than w_4. It can be seen how those agents that are more opened to alternatives are privileged. GPSao considers the openness to alternatives, the magnitude of the agents' preferences and the relationship between those preferences from a global perspective.

Once the mediator has applied the aggregation function, next step is to decide if negotiation ends or if a new set of alternatives C^{t+1} needs to be provided to the agents. This is the **contract selection** step.

2.4 Contract Selection

The starting point is the set Gp^t of group preferences and the set of contracts C^t. We associate with each contract c_i^t a probability

$$P(c_i^t) = \frac{(D_i^t)^\sigma}{\sum_{i=0}^{4}(D_i^t)^\sigma}.$$

The contract selection process selects the winner contract within C^t using a biased random experiment with these probabilities. The parameter $\sigma > 0$ works as an indication of the significance we give to the group preferences. If $\sigma \to \infty$ we select the contract with the maximum support. If $\sigma = 1$ then the probability of selecting

c_i^t would be proportional to its group support. The rationale behind using this probabilistic process is to introduce randomness and avoid local optima. However, this selection process it is not considering how good is the selection made.

The mediator must consider both the support values and their relationship to make the decision of expansion and contraction. Thus, we make σ vary as a function of D and the number of rounds t. If D is high, σ must be high, favouring a deterministic selection, i.e. with a high probability the contract with a higher D is selected. Otherwise, if D is low, σ must be low to induce randomness and avoid local optima. More specifically, for $\sigma = 0$ the selection of alternatives is equiprobable, making such selection independent of D. For $\sigma = 1$ the selection probability is proportional to D. Higher values for σ increases the probability of choosing the contract with a higher D.

To control σ we define the function:

$$\sigma(t, D) = \sigma_{min} + (\sigma_{max} - \sigma_{min}) \cdot D^{(1-\frac{t}{t_{max}})\cdot\alpha},$$

where σ depends on the negotiation round t, the maximum number of rounds t_{max} and D. The function is bounded by σ_{max} and σ_{min} given $D = 0$ and $D = 1$ respectively. The parameter $\alpha > 0$ determines the curvature of $\sigma(t, D)$. As the number of rounds t increases, the function increases its concaveness, which means that D induces higher values for σ, favouring convergence. Figure 3 shows the evolution of $\sigma(t, D)$

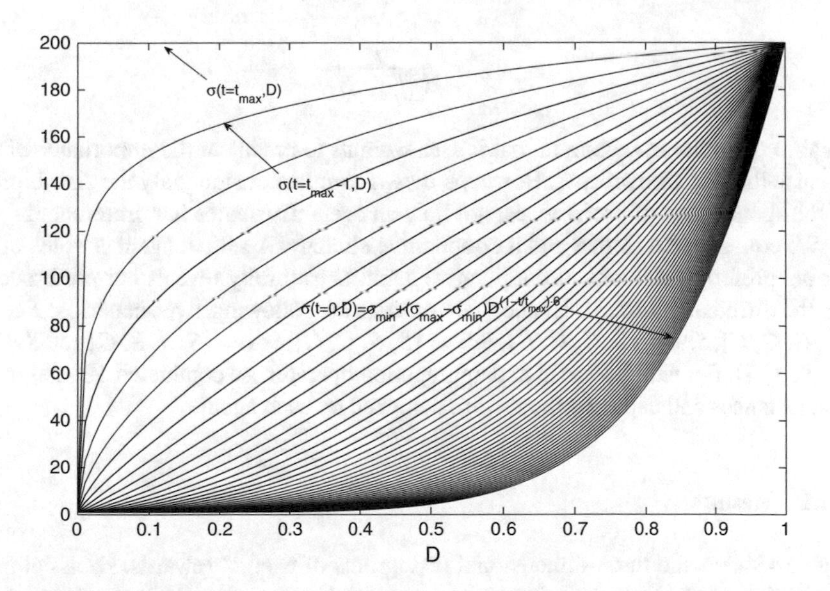

Fig. 3 Evolution of $\sigma(t, D)$ for $t_{max} = 50$, $\alpha = 6$, $\sigma_{max} = 200$ and $\sigma_{min} = 1$

for t_{max} =50, $\alpha = 6$, $\sigma_{max} = 200$ and $\sigma_{min} = 1$. The principle of this approach is analogous to the simulated annealing technique without reannealing. We can also introduce reannealing for $t_r < t_{max}$ such that t/t_{max} converts into $\frac{t-t_r}{t_{max}-t_r}$.

3 Experimental Evaluation

3.1 *Proof of Concept Scenario*

We evaluated a proof of concept scenario with four agents, contracts with two issues in the real domain, and reward functions which were built using an aggregation of *Bell functions*.

The configuration of parameters in the mediator was: $t_{max} = 50$ rounds, step-length threshold $1e-6$, and ($\alpha = 6, \sigma_{min} = 0, \sigma_{max} = 200$) for the selection process. We tested the performance of three different aggregation operators: the normalized sum operator (NSao), the Yager's operator (YAo) and GPSao. NSao sums up the agents' preferences. YAo applies the following formulas to aggregate preferences:

$$S_p = \sum_{i=1}^{n} A_{a_p}(c_i)$$

$$w_{a_p} = \frac{S_p}{\sum_{j=1}^{n} S_j}.$$

YAo is similar to GPSao in that it uses weights to modulate the importance of an agent in the selection process. However, these formulas consider only the distribution of the agents' sum of preferences, not how an agent distributes her preferences.

We considered a selfish and a cooperative strategy. A selfish agent S votes only for her preferred contract. A cooperative agent C truthfully reveals her preferences for the different contracts. Hence, we consider the following five scenarios: $Sc_1 = [C, C, C, C]$, $Sc_2 = [S, C, C, C]$, $Sc_3 = [S, S, C, C]$, $Sc_4 = [S, S, S, C]$ and $Sc_5 = [S, S, S, S]$. For each scenario and aggregation operator we conducted 100 negotiation instances and captured the reward achieved by each agent.

3.1.1 Results

Figure 4 shows the three-dimensional histograms of agents' rewards. Each column shows the set of histograms for a given aggregation operator. Each row represents one of the five behavior scenarios.

When all the agents are cooperative, independently of the aggregation operator, the group support is maximized (see first row). With NSao, selfish agents are clearly

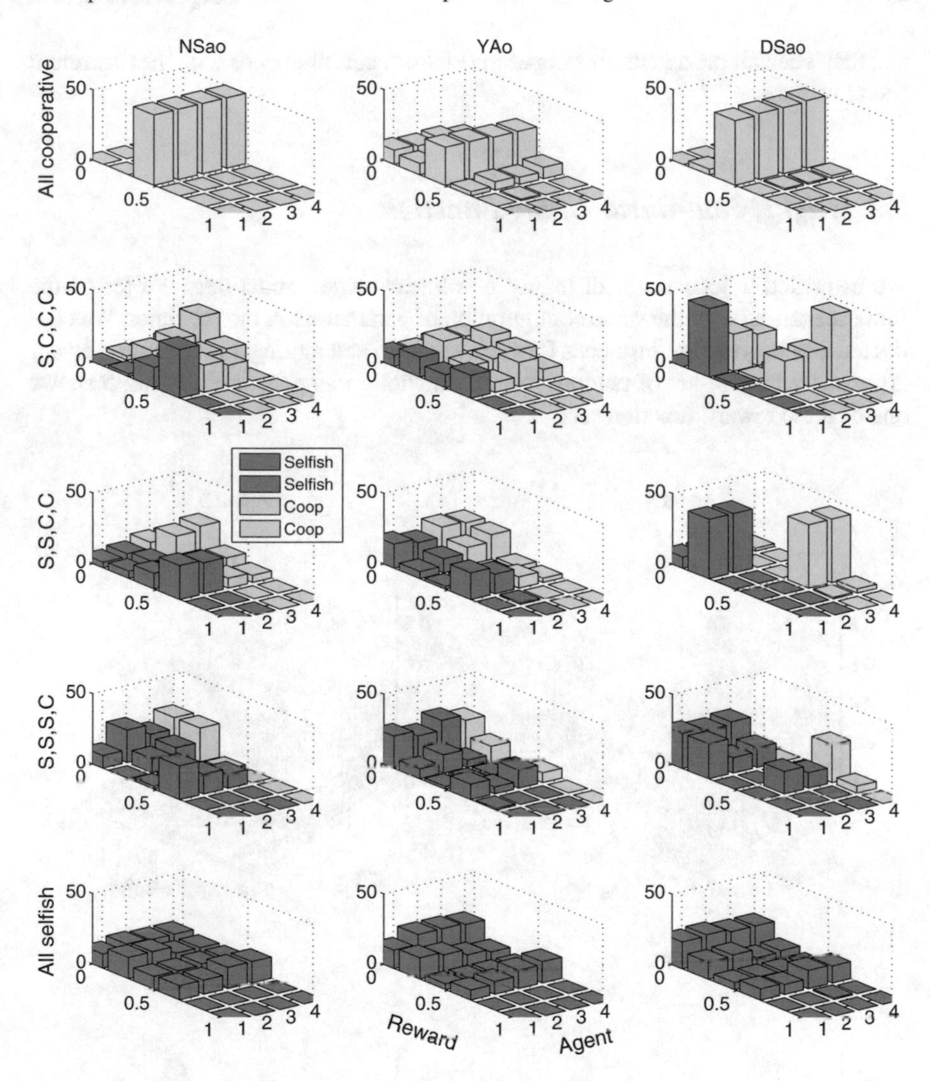

Fig. 4 Three-dimensional histograms of agents' rewards for the proof of concept scenario

privileged. However, if all the agents are selfish, the agents' rewards decrease drastically. NSao operator therefore does not provide a dominant strategy that maximizes social welfare. On the other hand, YAo operator performs even worse. The overall results when one or more selfish agents appear are very poor and negotiations exhibit a random behavior.

With GPSao, agents have an incentive to act cooperatively and truthfully reveal their preferences. We can see how selfish agents are expelled from the negotiation and that when all the agents are cooperative then negotiation ends with the maximum social welfare.

3.2 Highly Non-linear Utility Functions

We evaluated a scenario with highly non-linear reward functions. We tested the same scenarios, with the same configuration of parameters in the mediator. We conducted 100 negotiation instances for each scenario and aggregation operator. Every 20 instances a new set of random reward functions was generated. Figure 5 shows one of these reward function sets.

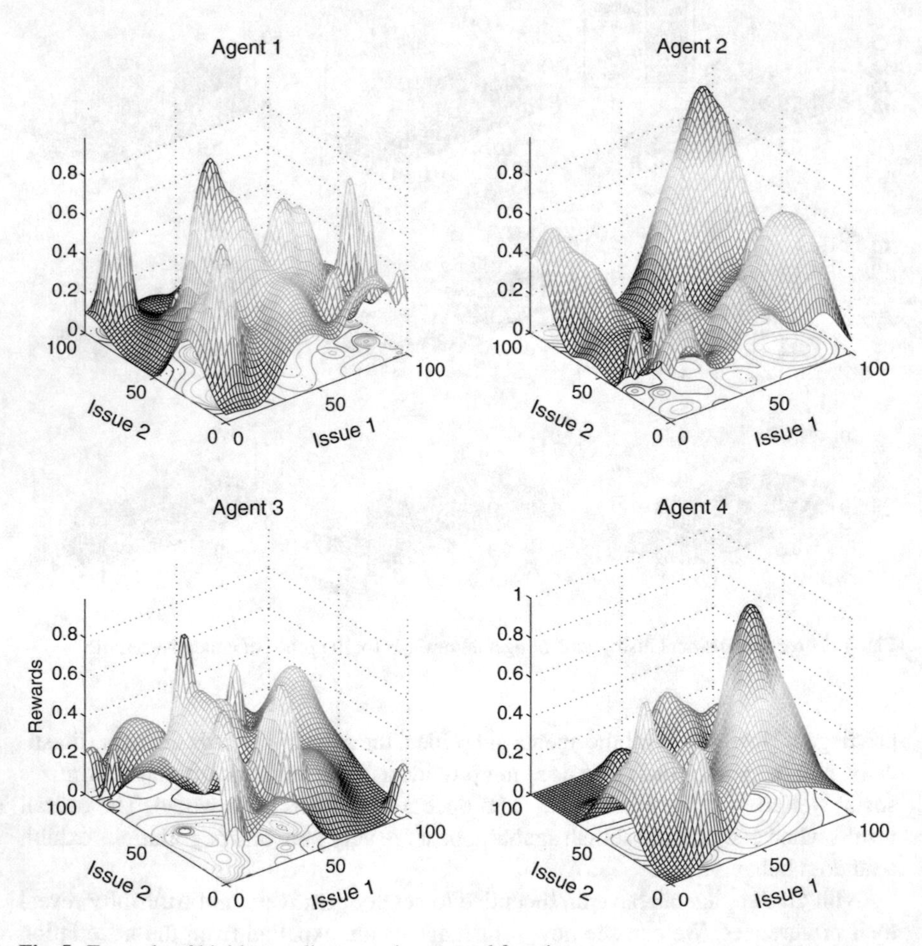

Fig. 5 Example of highly non-linear random reward functions

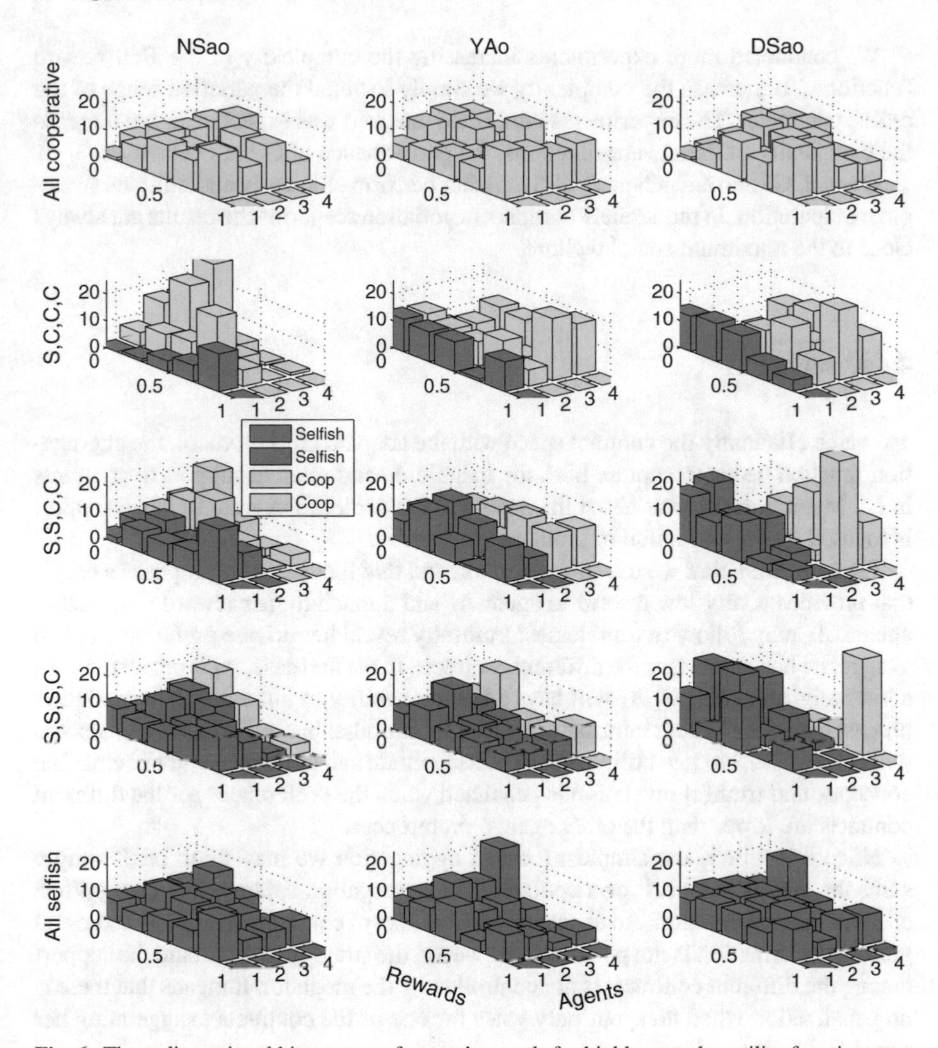

Fig. 6 Three-dimensional histograms of agents' rewards for highly complex utility functions

To generate the reward functions we varied randomly the center, height and radius parameters of the Bell reward function. The number of bells was fixed to 30.

3.2.1 Results

Figure 6 shows the 3D histograms of agents' rewards. As in the proof of concept scenario, with the GPSao operator agents have an incentive to act cooperatively and truthfully reveal their preferences. However, YAo operator exhibits a random behavior in all the scenarios, while NSao clearly privilege selfish agents.

We conducted more experiments increasing the complexity of the Bell reward functions. To increase the complexity we simply lowered the variation range of the radius parameter. The experiments confirmed that as it was expected, as we increase the complexity of the reward functions, the performance of GPSao decreases.

Overall, GPSao and adapted GPS together perform efficiently and mitigate strategic manipulation. In moderately complex negotiation scenarios the results are always close to the maximum social welfare.

4 Discussion

To search efficiently the contract space with the adapted GPS protocol, the aggregation function needs to capture both the magnitude and relative support to contracts in C^t. Magnitude informs about the quality of the exploration area. Relative support is focused on the extraction of gradient information.

Let us assume that we use YAo operator, and that the mediator proposes a set C^t that provides a very low reward to agent A_1 and a much higher reward to the other agents. A_1 may follow two strategies: truthfully reveal her private preferences, or to exaggerate her support for the different contracts. In the first case, A_1 is penalized with a low weight w_1. Hence, A_1 will have a low probability to influence the negotiation process because gradient information is lost. On the other hand, if A_1 exaggerates, she will conduct GPS to low utility contracts. Magnitude and gradient are not compatible concepts, and truthful revelation is penalized when the preferences for the different contracts are lower than the other agents' preferences.

NSao operator is the simplest form of aggregation we may think of. It simply sums the agents' support for a contract. The aggregation captures both magnitude and gradient information. An agent can support a set of contracts with a low score and gradient information is not penalized. However, the strategy to distribute the support among the different contracts is not controlled by the mediator. It means that there is no penalization when an agent only votes for one of the contracts exaggerating her support.

The main drawback of NSao is consequently that it does not provide a mechanism to control the distribution of scores. YAo controls the distribution at multi-agent level (i.e., it makes a relative comparison between the scores of the different agents), but it does not provide a mechanism to control how an agent distributes her support to the different contracts in C^t.

5 Conclusions

In this work we considered a mediated multiagent decision making problem. We proposed a multiagent negotiation framework where a mediator iteratively proposes a set of alternatives to a group of agents. Agents provide their support to the different alternatives, and then the mediator evaluates the group support to the different alternatives. We proposed the GPSao aggregation operator that considers both the magnitude and distribution of the support to the alternatives in order to make a fair aggregation of preferences. The mediator applies an adapted optimization algorithm (GPS) to generate the different sets of alternatives, being an objective to obtain solutions close to the maximum social welfare. We have evaluated GPSao against the Normalized Sum and the Yager's aggregation operators [19]. Yager's operator avoids strategic manipulation in scenarios of one-shot group decision making but fails in multi-shot scenarios. The experimental evaluation shows that our negotiation framework and GPSao mitigates strategic manipulation both in one-shot and multi-shot group decision making and find solutions close to the maximum social welfare.

The proposed framework is limited to alternatives in the real domain, with preference functions that exhibit a medium correlation distance, i.e., they can be non-monotonic but must have smooth shapes. Agents with very restrictive preference functions may be truthfully revealing their preferences but be expelled from the negotiation. Also, if we change to domains where alternatives are defined in terms of categories or discrete domains, adapted GPS and the GPSao may not work. In these cases we will need to reformulate both GPS and GPSao, or use other optimization mechanisms better adapted to these domains. We need to explore how to cope with more complex reward function types. Finally, we have assumed that agents do not know the preferences of the other agents. In our opinion, we need to explore the situations where agents may have partial knowledge. Basically, we need to assess if such knowledge may cause strategic manipulation.

Acknowledgements This work has been supported by the Spanish Ministry of Economy and Competitiveness grant: TEC2013-45183-R CIVTRAff.

References

1. T. Calvo, G. Mayor, R. Mesiar, *Aggregation Operators: New Trends and Applications*, vol. 97 (Springer, New York, 2002)
2. E. de la Hoz, M.A. López-Carmona, M. Klein, I. Marsá-Maestre, Consensus policy based multi-agent negotiation, in *Proceedings of the Agents in Principle, Agents in Practice - 14th International Conference, PRIMA 2011*, Wollongong, Australia, 16–18 November 2011, pp. 159–173
3. E. de la Hoz, M.A. López-Carmona, M. Klein, I. Marsá-Maestre, Hierarchical clustering and linguistic mediation rules for multiagent negotiation, in *International Conference on Autonomous Agents and Multiagent Systems, AAMAS 2012*, Valencia, Spain, 4–8 June 2012 (3 Volumes), pp. 1259–1260 http://dl.acm.org/citation.cfm?id=2343952
4. E. de la Hoz, M.A. López-Carmona, M. Klein, I. Marsá-Maestre, Consortium formation using a consensus policy based negotiation framework, *Complex Automated Negotiations: Theories, Models, and Software Competitions* (Springer, Berlin, 2013), pp. 3–22
5. H. Ehtamo, R.P. Hamalainen, P. Heiskanen, J. Teich, M. Verkama, S. Zionts, Generating pareto solutions in a two-party setting: constraint proposal methods. Manag. Sci. **45**(12), 1697–1709 (1999)

6. P. Heiskanen, H. Ehtamo, R.P. Hamalainen, Constraint proposal method for computing Pareto solutions in multi-party negotiations. Eur. J. Oper. Res. **133**(1), 44–61 (2001)
7. T. Ito, M. Klein, H. Hattori, A multi-issue negotiation protocol among agents with nonlinear utility functions. J. Multiagent Grid Syst. **4**(1), 67–83 (2008)
8. M. Klein, P. Faratin, H. Sayama, Y. Bar-Yam, Protocols for negotiating complex contracts. IEEE Intell. Syst. **18**(6), 32–38 (2003)
9. G. Lai, K. Sycara, A generic framework for automated multi-attribute negotiation. Group Decis. Negot. **18**, 169–187 (2009)
10. F. Lang, A. Fink, Learning from the metaheuristics: protocols for automated negotiations. Group Decis. Negot. **24**(2), 299–332 (2015). doi:10.1007/s10726-014-9390-x
11. R.M. Lewis, V. Torczon, M.W. Trosset, Direct search methods: then and now. J. Comput. Appl. Math. **124**, 191–207 (2000)
12. M. Li, Q.B. Vo, R. Kowalczyk, Searching for fair joint gains in agent-based negotiation, in *Proceedings of the 8th International Conference on Autonomous Agents and Multiagent Systems (AAMAS 2009)*, ed. By S. Decker, C. Sierra, Budapest, Hungary (2009), pp. 1049–1056
13. M.A. Lopez-Carmona, I. Marsa-Maestre, E. de la Hoz, J.R. Velasco, A region-based multi-issue negotiation protocol for non-monotonic utility spaces, in *Computational Intelligence* (In press), pp. 1–48 (2011), doi:10.1007/s10458-010-9159-9
14. M.A. Lopez-Carmona, I. Marsa-Maestre, M. Klein, T. Ito, Addressing stability issues in mediated complex contract negotiations for constraint-based, non-monotonic utility spaces. J. Auton. Agents Multiagent Syst. 1–51 (2010) (Published online)
15. I. Marsa-Maestre, M.A. Lopez-Carmona, J.R. Velasco, T. Ito, M. Klein, K. Fujita, Balancing utility and deal probability for auction-based negotiations in highly nonlinear utility spaces, in *21st International Joint Conference on Artificial Intelligence (IJCAI 2009)*, Pasadena, California, USA (2009), pp. 214–219
16. D. Pelta, R. Yager, Analyzing the robustness of decision strategies in multiagent decision making. Group Decis. Negot. **23**(6), 1403–1416 (2014). doi:10.1007/s10726-013-9376-0
17. D.A. Pelta, R.R. Yager, Decision strategies in mediated multiagent negotiations: an optimization approach. IEEE Trans. Syst. Man Cybern. Part A **40**(3), 635–640 (2010). doi:10.1109/TSMCA.2009.2036932
18. Q.B. Vo, L. Padgham, L. Cavedon, Negotiating flexible agreements by combining distributive and integrative negotiation. Intell. Decis. Technol. **1**(1–2), 33–47 (2007)
19. R. Yager, Multi-agent negotiation using linguistically expressed mediation rules. Group Decis. Negot. **16**(1), 1–23 (2007)
20. R. Yager, J. Kacprzyk, *The Ordered Weighted Averaging Operators: Theory and Applications* (Kluwer, Boston, 1997)

A Dependency-Based Mediation Mechanism for Complex Negotiations

Akiyuki Mori, Shota Morii and Takayuki Ito

Abstract There has been an increasing interest in automated negotiation and particularly negotiations that involves multiple interdependent issues, which yield complex nonlinear utility spaces. However, none of the proposed models were able to find a high-quality solution within a realistic time. In this paper we presents a dependency-based mediation mechanism for complex negotiations. In the complex negotiation field, there have been works on issue-by-issue negotiation models. When considering a real world negotiation with a mediator, a mediator proposes issues, and players negotiate on those issues.Then if there is no agreement, the other issues are proposed by the mediator so that players can find a possible agreement. The sequence of proposing issues should be based on the dependency of those issues. In this paper, we adopt a dependency-based complex utility model (Hadfi and Ito, Modeling decisions for artificial intelligence (2014) [12]), which allow a modular decomposition of the issues and the constraints by mapping the utility space into an issue-constraint hypergraph. Based on this model, we propose a new mediation model that can efficiently mediate an issue-by-issue negotiation while keeping privacy on the utility values. Our experimental results show our new mediation model can find agreement points that are close to Pareto front in efficient time periods.

Keywords Automated multi-issue negotiation · Dependency-based mediation mechanism · Interdepend issues · Constraint-based utility spaces · Hyper-graph

---

A. Mori (✉) · S. Morii · T. Ito
Nagoya Institute of Technology, Aichi, Japan
e-mail: mori.akiyuki@itolab.nitech.ac.jp

S. Morii
e-mail: morii.shouta@itolab.nitech.ac.jp

T. Ito
e-mail: ito.takayuki@nitech.ac.jp

© Springer International Publishing AG 2017

K. Fujita et al. (eds.), *Modern Approaches to Agent-based Complex Automated Negotiation*, Studies in Computational Intelligence 674,
DOI 10.1007/978-3-319-51563-2_4

1 Introduction

Negotiation is an important process in making alliances and represents a principal topic in the field of multi-agent system research. There has been extensive research in the area of automated negotiating agents [5, 17, 21, 22, 24, 25]. Automated agents can be used side-by-side with a human negotiator embarking on an important negotiation task. They can alleviate some of the effort required of people during negotiations and also assist people who are less qualified in the negotiation process. There may even be situations in which automated negotiators can replace the human negotiators. Thus, success in developing an automated agent with negotiation capabilities has great advantages and implications.

In particular, for automated negotiation agents in bilateral multi-issue closed negotiations, attention has focused on forming agreements with negotiation strategies [1, 2, 4]. Many existing studies postulate the utility of the issues independence, as a result the utility of the agent could be expressed as a linear utility function. However, when considering negotiation problems in the real world, it is not often that each issue is independent. In the other words, the issues are dependent on each other. Furthermore, Klein et al. show that the technique which can find good agreement point in the utility space of the independence issues does not work effectively if the issues are interdependent [20]. Therefore, this paper discusses the complex negotiation problems in an interdependent relationship, which a more realistic and the amount of computation is enormous.

Motivated by the challenges of bilateral negotiations between people and automated agents, the Automated Negotiating Agents Competition (ANAC) was organized [10, 15, 16, 18, 31]. The purpose of the competition is to facilitate research in the area of bilateral multi-issue closed negotiation, and especially the setup at ANAC2014 adopted complex negotiation problems in multiple interdependence issues using nonlinear utility function.

In complex nonlinear utility spaces, One of the main challenges is scalability problem; they cannot find a adequate solutions when there are a lot of issues, due to computational intractability. Our reasonable approach to reducing computational cost, while maintaining good quality outcomes, is to negotiate the issues sequentially, which is a representative game theoretic models called issue-by-issue negotiation [3, 7, 13]. Although issue-by-issue negotiation reduces the computational cost, such approaches do not consider the negotiation order in many issues. For Instance, if there are two issues, X and Y, an important question that arises is how to determine the first negotiation issue. In other words, the two order XY and YX can result two different outcomes. Therefore, they need to determine the order of issues, but exploring the optimal issue is a difficult task, especially given that real world negotiations involve multiple issues. For these reason, in the complex multiple issues, it is desirable considering dependency among the issues.

Hadfi et al. proposed a modular representation for nonlinear utility spaces by decomposing the constraints and issues into an utility hypergraph [12]. Exploration and searching for optimal contracts are performed based on a message-passing mechanism in the hypergraph. This mechanism can handle a large family of complex utility spaces by finding the optimal contracts, outperforming previous sampling-based approaches. However, this model which finds for the solutions in complex interdependent spaces are confined to explore for the own maximum solutions in the utility space. Hence, this model do not discuss a methods that search and negotiate the best agreement points in the whole of negotiation participants.

In this paper, we propose a new negotiation model based on the dependency-based mediation mechanism in complex nonlinear utility spaces. Specifically, we proposed method for reaches an optimal consensus each other by means of representing the preference information of the agents as a hypergraph. In addition to, we adopt a mediator to be fair and facilitate negotiation. We propose a mediator strategy that expands sequentially the negotiation issues at an optimal time so that agents allow for smooth negotiation. Note that this approach that mediator expand the issues gradually is well-known in the real world negotiations. We experimentally evaluated our model using several nonlinear utility spaces, showing that it can handle large and complex spaces by finding the good solutions in the minimum time.

The remainder of this paper is organized as follows. In the next section, we propose the basics of nonlinear utility space representation. In Sect. 3, we describe our new model based on a dependency-based mediation mechanism for optimal contracts search. In Sect. 4, we provide our experimental results, and in Sect. 5, we describe related works. In Sect. 6, we conclude and outline future work.

2 Negotiation Environments

2.1 Multi-issue Automated Negotiations

In the field of multi-issue automated negotiation, the **bilateral negotiation protocol** is one of the most popular protocols while mediator-based negotiation protocols have been proposed as well. In a bilateral negotiation protocol, two agents directly exchange their offers to make an agreement. One well-known bilateral negotiation protocol is the alternating-offers protocol proposed by Rubinstein [27–29]. This protocol has been adopted as the standard protocol in the Automated Negotiating Agent Competitions (ANAC). The purpose of this competition is to steer research in the area of bilateral multi-issue closed negotiation. In closed negotiations, when opponents do not reveal their preferences to each other, which is an important class of real-life negotiation. Negotiating agents designed using a heuristic approach require extensive evaluation, typically through simulations and empirical analysis, since it is usually impossible to predict precisely how the system and constituent agents will behave in a wide variety of circumstances [19, 30].

The following shows an example of the alternative-offers protocol in ANAC. Let us think about agent A and agent B. First, agent A proposes an offer (a bid) to agent B. Then, agent B can choose the following three actions based on the offer proposed by agent A: *Accept* embraces the offer by the opponent, and the offer becomes the agreement. Agents get utilities based on the agreement. *Offer* rejects the offer and then proposes a new offer to the opponent. Negotiation continues. *EndNegotiation* gives up the entire negotiation. If one of agents chooses this option, the negotiation ends without agreement. Thus, the utilities that agents can get is the minimum or 0.

In a **mediator-based negotiation**, a mediator (or mediators) is supposed to exist between the negotiating agents. The mediator is usually assumed to be fair and facilitate negotiation so that agents can reach agreements and obtain higher utilities. Mediator-based negotiation can also be used with bilateral negotiation while a mediator is used for managing multi-party negotiations.

2.2 Nonlinear Utility Models

In the field of complex negotiations, nonlinear utility is assumed in the sense that the shape of the utility function is not monotonic, rather bumpy. One common nonlinear utility model is the multi-issue interdependent utility model, where we assume multiple issues are depended on each other. Such interdependency causes non-monotonic and bumpy shapes. Actually, it would be possible to assume that the real human utility function is more non-monotonic.

Constraint-Based Utility Model:

The classical and widely used multiple interdependent issue utility model is the constraint-based utility model [8, 14], which is defined by the accumulation of constraint blocks. A constraint is constrained with several issues, and if these issues are satisfied, an agent can obtain utility from this constraint. For example, let us assume there are issues, "color", "type", "engine", "power", etc., for buying a car. One constraint could represent that if "color" is "red" or "blue" and "type" is "sports", then the agent has utility 50 for this car. We assume an agent has a lot of these constraint blocks. Accumulation of these constraint blocks make nonlinear utility space.

Formally, assume M issues and each issue $i_j \in I$ (I is the set of issues) has issue value $s_j \in [0, X]$, where $[0, X]$ is the domain of issue i_j. A possible contract is represented as vector $\mathbf{s} = (s_1, s_2, \dots, s_M)$. Let us represent a constraint as $\varphi_k \in \Phi$ (Φ is the set of constraints). φ_k has a range of values for some of the issues, which satisfy φ_k. For example, the following representation is possible that c_1 is satisfied and has utility 300 if i_1 is in $[3, 5]$ and i_2 is in $[4, 8]$. Constraint φ_k has a utility $w(\varphi_k, \mathbf{s})$ when $\mathbf{s}$ satisfies φ_k.

Agent $n \in [0, 1]$'s utility for possible contract $\mathbf{s}$ can be defined as the following Eq. (1).

$$u_n(\mathbf{s}) = \sum_{\varphi_k \in \Phi, \mathbf{s} \in x(\varphi_k)} w_n(\varphi_k, \mathbf{s}), \quad (1)$$

where $x(\varphi_k)$ is a set of possible contracts that can satisfy φ_k.

This constraint-based utility function representation allows us to capture the issue interdependencies common in real world negotiations. A negotiation protocol for complex contracts can, therefore, handle linear contract negotiations. The objective function for our protocol can be described by the following Eq. (2).

$$\arg\max_{\mathbf{s}} \sum_{n \in [0,1]} u_n(\mathbf{s}). \tag{2}$$

Our protocol, in other words, tries to find contracts that maximize the *social welfare* (i.e., the total utilities for all agents) within a realistic time. Such contracts, by definition, will also be Pareto-optimal.

Figure 1 shows an example of a utility space generated via a collection of binary constraints involving issues i_1 and i_2. The utility function is highly nonlinear with many hills and valleys. Having a large number of constraints produces a *bumpy* nonlinear utility space with high points whenever many constraints are satisfied and lower points where few or no constraints are satisfied. This representation is intuitive and simple to implement. Thus there have been a lot of studies based on this model. However, because this model utilizes a multi-dimensional numerical graph, it is not

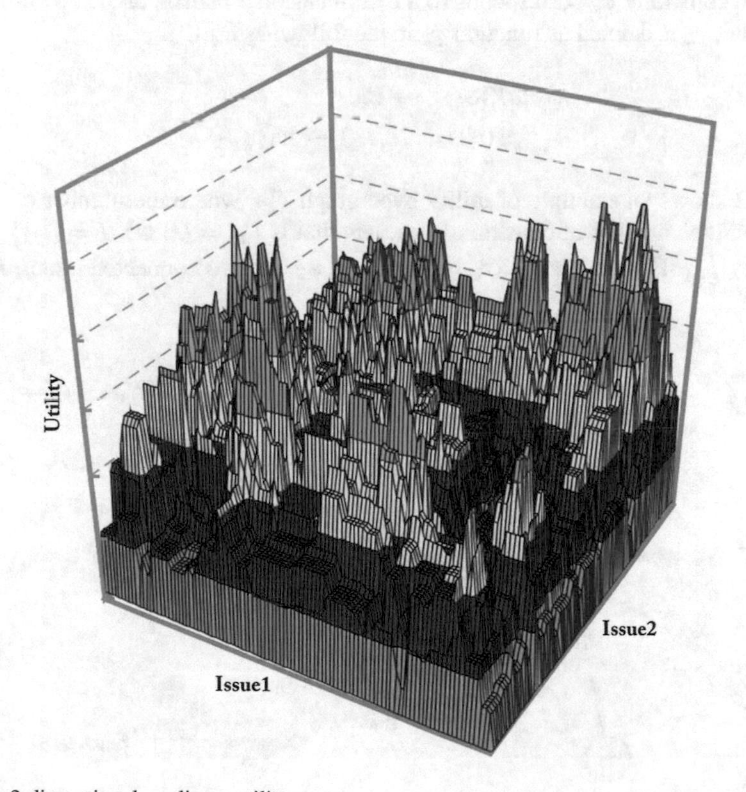

Fig. 1 2-dimensional nonlinear utility space

natural to think about discrete and uncomparable values of issues. On the other hand, there have been several approaches to use graph-structures to represent the nonlinearity of utility space. The following dependency-based hypergraphical utility model is one recent nonlinear utility space model.

2.3 Dependency-Based Hypergraphical Utility Model

To represent complex utilities, there have been several approaches that utilize graph structures [9, 26]. In this paper we focus on the dependency-based hypergraphical utility model [12] because this model can represent the other graph based structures and also the above constraint-based utility model.

Let us define utility hypergraph $\Gamma = (I, \Phi)$, with sets of issues I and constraints Φ. $\Phi = \{\varphi_k\}_{k=1}^{l}$, where φ_k is the k-th constraint and l is the total number of constraints.

To each constraint $\varphi_k \in \Phi$, this model assigns neighbors set $\mathcal{N}(\varphi_k) \subset I$ that contains issues connected to φ_k with $|\mathcal{N}(\varphi_k)| = \delta_k$

Each constraint φ_k corresponds to a δ_j-dimensional matrix, $\mathcal{M}_{\varphi_k}$. The utility of constraint φ_k is defined as function ϕ_k in the following Eq. (3).

$$\begin{aligned} \phi_k :& \mathcal{N}(\varphi_k)^{\delta_k} \to \mathbb{R} \\ & \phi_k(i_1, i_2, \ldots, i_{\delta_k}) \mapsto w(\varphi_k, \mathbf{s}). \end{aligned} \tag{3}$$

Figure 2 shows an example of utility hypergraph Γ_{10}, where the number of issues is ten and the number of constraints is seven, that is $\Gamma_{10} = (I, \Phi)$, $I = \{i_j\}_{j=0}^{9}$ and $\Phi = \{\varphi_k\}_{k=1}^{7}$. Here, $\mathcal{N}(\varphi_3) = \delta_2 = 2$ because φ_3 has two connected issues, i_1 and

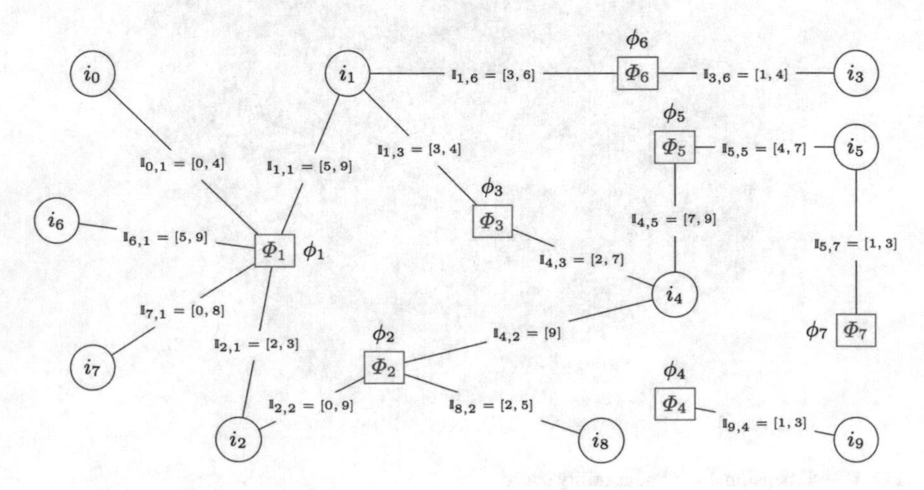

Fig. 2 Hypergraph Γ_{10} consists of issues set I and constraints set Φ

i_4, where φ_3 is satisfied if the value of i_1 is in [3, 4] and the value of i_4 is in [2, 7]. That is, if $\mathbf{s} = (3, 2, 2, 4, 5, 4, 3, 1, 2, 3)$, φ_3 is satisfied, and then the agent obtains utility $w(\varphi_3, \mathbf{s})$.

3 A Dependency-Based Mediation Mechanism

3.1 Negotiation Protocol

This paper proposes a 2-agent mediated negotiation protocol based on the hypergraph utility model. In this protocol, agents do not need to open their utility-value information (called *preference profiles*) while revealing only its hypergraph structure information. The mediator can efficiently facilitate negotiations using the shared hypergraph structure information. The following show our proposed negotiation model and show the negotiation steps in Fig. 3.

Step1: Submission and generation of hypergraphs

Let $\mathscr{D}$ be our set of domains. For every domain $D \in \mathscr{D}$ two preference profiles exist, $\mathscr{P}^D = \{P_1^D, P_2^D\}$. Suppose agent A negotiates with B in domain $D \in \mathscr{D}$, where A has the first preferences profile P_1^D and B uses P_2^D. Agents A and B generate two hypergraphs, $\Gamma_A = (I_A, \Phi_A)$ and $\Gamma_B = (I_B, \Phi_B)$ based on their own preference profile. Note that a set of issues is $I_A = I_B$ (namely, $I_A \subset I_B$ and $I_B \subset I_A$), but a set

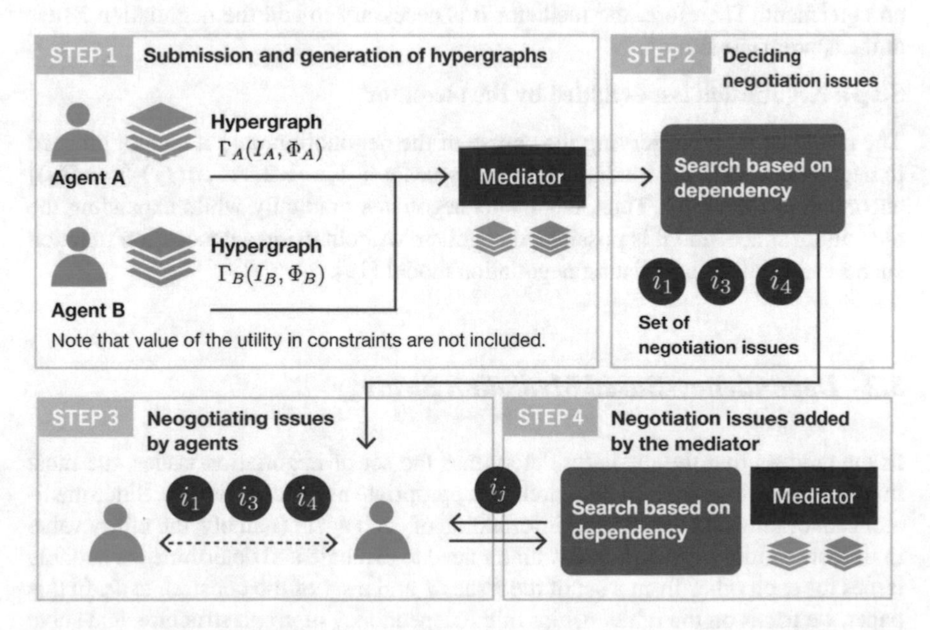

Fig. 3 Process of dependency-based mediation mechanism

of constraints is not always $\Phi_A = \Phi_B$, generally $\Phi_A \neq \Phi_B$ in many utility spaces. Similarly, the number of a set of constraints $n(\Phi)$ is $n(\Phi_A) \neq n(\Phi_B)$ in general.

Each agent submits the generated hypergraph to the mediator, and the mediator obtain the information of hypergraph of each agent. Note that in the information of hypergraph submitted, agent's preference profile, that is, the value of the utility in constraints $w(\varphi_k, \mathbf{s})$, are not included. Therefore, information the mediator can know is a graph structure in issues set I and constraint set Φ only.

Step2: Deciding negotiation issues by mediator

Based on the submitted hypergraph A and B by *Step1*, the mediator decides $n(>0)$ as the number of the current negotiation issues. In addition to, let us define a set of negotiation issues that is presented by the mediator as $I_m \subset I_A (= I_B)$. Negotiation issues are determined from the dependency of the graph structure, and the mediator presents the issues where both agents are expected to be able to proceed negotiation smoothly.

Step3: Negotiating issues by agents

Agents negotiate on the negotiation issues that are presented by the mediator in *Step2*. Each agent negotiate in the set of issues I_m, and compromise at one's discretion. For the set of issues except the set I_m, the agent select the values to maximize own utility. As a result, it is possible to negotiate high priority issues for each other by the agents who negotiate in the utility space of the set of issues I_m.

However, since the number of negotiation issues is small as well as utility space is narrow, negotiation is not performed well, and it is often the case that can not reach an agreement. Therefore, the mediator it is necessary to add the negotiation issues at the appropriate timing.

Step4: Negotiation issues added by the mediator

The mediator while observing the course of the negotiation, and if cannot proceed to negotiation well, the mediator add the issues $n+1, n+2, \ldots, n(I_A)$ $(= n(I_B))$ according to the *Step2*. Thus, the agents negotiates gradually while expanding the own utility space, and it is possible to negotiate without using a concession function such a essential for an existing negotiation model [19].

3.2 Dependency-Based Mediation Strategy

In the process that the mediator determines the set of negotiation issues, the most important is how to decide and search the appropriate negotiation issues. Since mediator cannot know the preference information of each agent (namely, the utility value in the constraints $w(\varphi_k, \mathbf{s})$), the mediator need to evaluate and determine the suitable issues for each other from a set of the issues I and a set of the constraints Φ. In this paper, we focus on the relationships of the dependency of graph structure, and solve these problems by exploring the issues that both agents has common dependencies.

Algorithm 1 Setting Negotiation Issues I_m

Require: $k > 0$ {k is the number of current negotiation issues} i_{random} {$i_{random} \in I$ is a random issue}
Ensure: I_m
1: **if** $I_m = \emptyset$ **then**
2: $I_m \leftarrow I_m \cup \{\psi(i_{random})\}$
3: $I_e \leftarrow I_e \cup i_{random}$
4: **end if**
5: **while** $n(I_m) < k$ **do**
6: $i_e \leftarrow \max_{i_k}(\{n(\nu_{AB}(i_j, i_k)) \mid i_j \in I_m \wedge i_j \notin I_e \wedge i_k \in \psi(i_j)\})$
7: **if** $i_e = \emptyset$ **then**
8: $i_e \leftarrow i_{random} \notin I_m$
9: **end if**
10: $I_m \leftarrow I_m \cup i_e$
11: $I_e \leftarrow I_e \cup i_e$
12: **end while**

We describe the search strategy of the mediator based on the dependency of the graph structure as follows, and we show the algorithm (see Algorithm 1)

First, we define $\nu(i_j)$, which the issue i_j have the set of the constrains, that is, $\nu(i_j) = \{\varphi \in \Phi \mid i_j \in \varphi\}$. Thereby, the relationships of the dependencies are represented as a set of the issues that including every issues (excluding i_j) connected by $\nu(i_j)$. In other words, when we define the set of issues $\mathcal{N}(\varphi_k)$ that connected by φ_k, the dependencies are represented in the following Eq. (4).

$$\psi(i_j) = \{i \in I \mid i \in \mathcal{N}(\nu(i_j)) \cap i \neq i_j\}. \tag{4}$$

The decisions of negotiation issues I_m are based on dependencies $\psi_A(i_j)$ and $\psi_B(i_j)$, and the mediator defines the set of mutual dependency issues in the following Eq. (5):

$$\begin{aligned} \psi_{AB}(i_j) &= \psi_A(i_j) \cap \psi_B(i_j) \\ &= \{i \in I \mid i \in \psi_A(i_j) \wedge i \in \psi_B(i_j)\}. \end{aligned} \tag{5}$$

The mediator selects the set of issues from $\psi_{AB}(i_j)$ (lines 1–4 in Algorithm 1). Note that it chooses the first issue randomly since prevent the agents from reaching localized solutions. The determined set of issues has the same dependency in both agents A and B, moreover, the graph structure for the issues is identical. As a result, it is possible to extract an important issues from each other, and the agents can conduct smooth negotiations.

Although the mediator can derive the set of issues I_m by the process above, it is necessary to further search when increasing the negotiation issues. For example, assume issue i_1 is obtained dependency $\psi_{AB}(i_1) = \{i_2, i_3\}$. At this time, set of issues I_m is $I_m = \{i_1, i_2, i_3\}$, when presenting four or more of the negotiation issues, the mediator needs to further carry out the search. However, there is a one problem that the mediator must choose the next searching issue from i_2 and i_3, and in order to extract an important issue for each other, these issues $\{i_2, i_3\}$ need to evaluate. Therefore, we define the number of dependency in the hypergraph, and the mediator search an issues based on the number of dependency.

The number of dependency is defined as the number of dependencies $n(\psi(i_j))$. The mediator calculate the sum of $n(\psi_A(i_j))$ and $n(\psi_B(i_j))$ (namely, $n(\psi_{AB}(i_j)) = n(\psi_A(i_j)) + n(\psi_B(i_j))$) which the number of dependencies from each agent, and the mediator set the next search issue having more larger dependencies (line 6, 10–11 in Algorithm 1). Therefore, from set of current negotiation issues I_m, calculate the following Eq. (6).

$$\max_{i_k} \left(\{n(\nu_{AB}(i_j, i_k)) \mid i_j \in I_m \wedge i_j \notin I_e \wedge i_k \in \psi(i_j)\}\right), \tag{6}$$

where I_e is a set of issues which have been searched already.

In the case of the number of dependencies is large, we can expect this issue to have many constraints. In the other words, we consider the more important issue. In addition to, the issue that the number of dependencies is largely connected to many other issues. Hence, it can be said to be an issue that need to search preferentially.

The mediator also defines the k parameter from the maximum dependencies and adds an extra issue without independence if she cannot find a new issue (lines 7–9 in Algorithm 1). This algorithm involves issues that are not shared by all the agents, but they can conduct smooth negotiations since at the beginning the mediator offers a set of issues with identical dependency.

4 Experimental Results

4.1 Settings

Next we examined whether our proposed model, that is, a mediated negotiation protocol based on the hypergraph utility model effectively working. In this experiment, we investigated by the comparison simulation between the existing model and the proposed model, and we show the usefulness of our model.

In each experiment, we ran 100 negotiations. The following parameters were used. The negotiation time was 180 s in the real time, but the time line is normalized, i.e.: time $t \in [0, 1]$, where $t = 0$ represents the start of the negotiation and $t = 1$ represents the deadline. The domain for the issue value was [0,9]. The number of issues was 10, 30, and 50 issues (These domain was used in the final of ANAC2014). Therefore, the number of bids was $10^{Number of Issues}$ (for instance, 10 issues domain produces a space of 10^{10} (=10 billion) possible contracts). Constraints that satisfy many issues have, on average, larger utility, which seems reasonable for many domains. The number of agents is 4 (Our Agent, *Gangster*, *WhaleAgent*, *E2Agent*) in these experiments, which we choose agents for comparison from advance to the finals in ANAC2014. We show the evaluation indexes in the comparative experiment below.

- Social Welfare: An average of the total of acquired utility for each other.
- Number of Bid: An average the number of *Bids*
- Accept Time: An average of the time that reach the *Accept*.

4.2 Discussion

Figures 4, 5, and 6 show the average of social welfare, the average of bids, and the average of accept time among the agents at each negotiation. The horizontal axis shows the acquired utilities (Fig. 4), the number of bids (Fig. 5), and the accept time (Fig. 6). The vertical axis indicates the kind of domains (namely, 10, 30, and 50 issues). Note that the accept time t is normalized to $0 \leq t \leq 1$.

Firstly, the result in Figs. 5 and 6 indicates that the proposed model can reach an agreement by the short negotiation time and the small number of bids. In the existing models, it was difficult to reach an agreement at short time or little bid, and

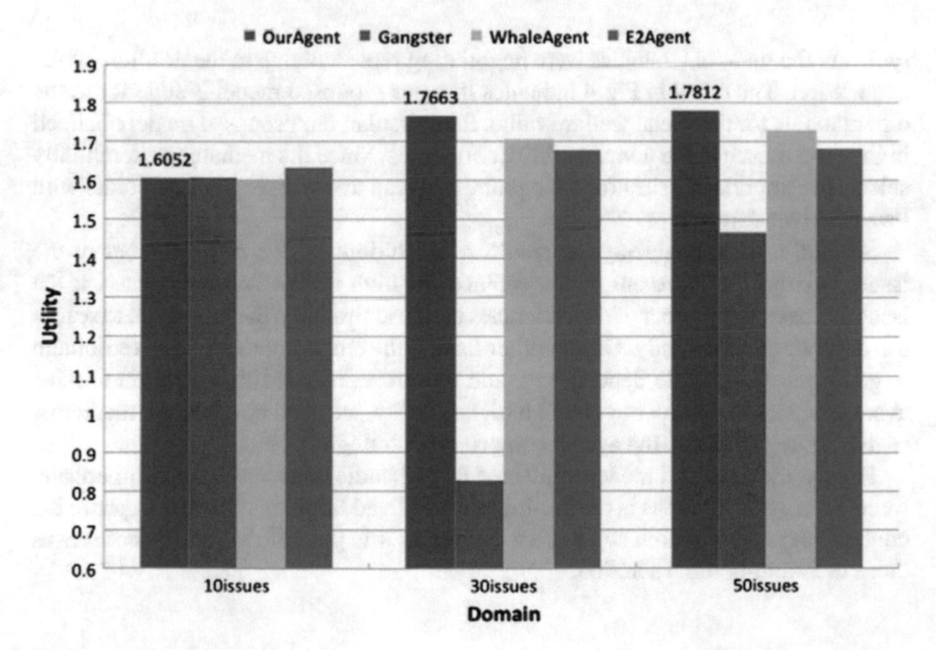

Fig. 4 Average of social welfare

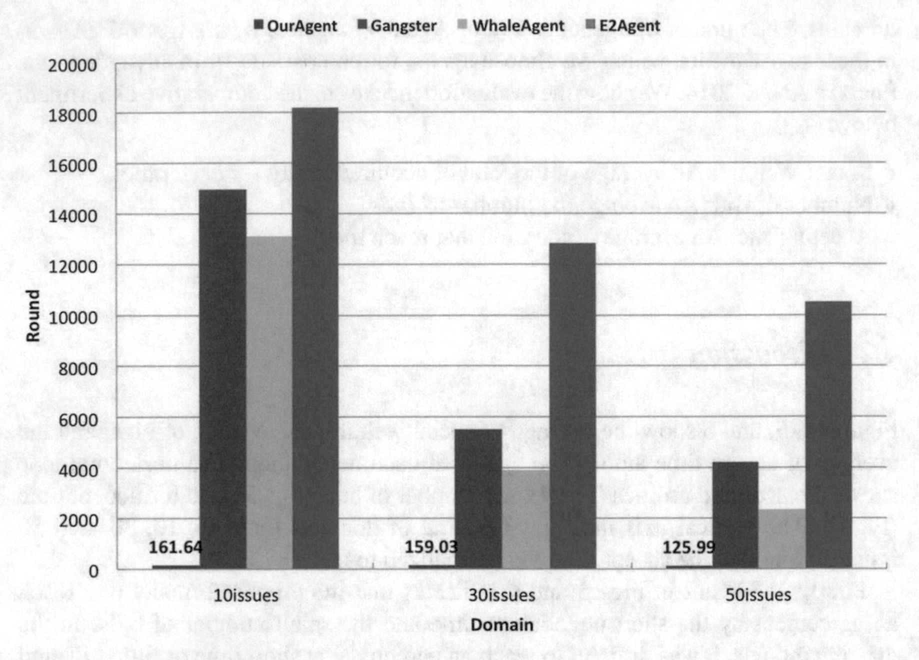

Fig. 5 Average number of bids

typically the main of strategies were negotiating repeatedly up to the deadline of the negotiation. The result in Fig. 4 indicates that our proposed model is superior to the other models for the social welfare value. In particular, our proposed model obtained high performance in the domain of 30 or 50 issues. Since the mediator preferentially selects the important issues for the agents, they can avoid agreement on points with low social welfare values.

Secondly, when analyzing the results of each domain, the more number of the issues is large, the more our model can acquire high utility. This is because, if the issue increase, the number of dependencies increase similarly, that is, enable to search the dependent issue easily. On the other hands, the small number of issues domain (e.g. 10issues) has little dependency, and therefore, reduced the acquired utilities averagely. Hence, in order to obtain a higher utility, we need not only the mediators search strategy but also the agents compromise strategies.

Finally, the proposed model confirmed it can handle large and complex spaces by finding the good solutions in the minimum time. In addition to, if further improve the concession and the search strategy of the agent, it is possible to reach a consensus close to Pareto-optimal solution.

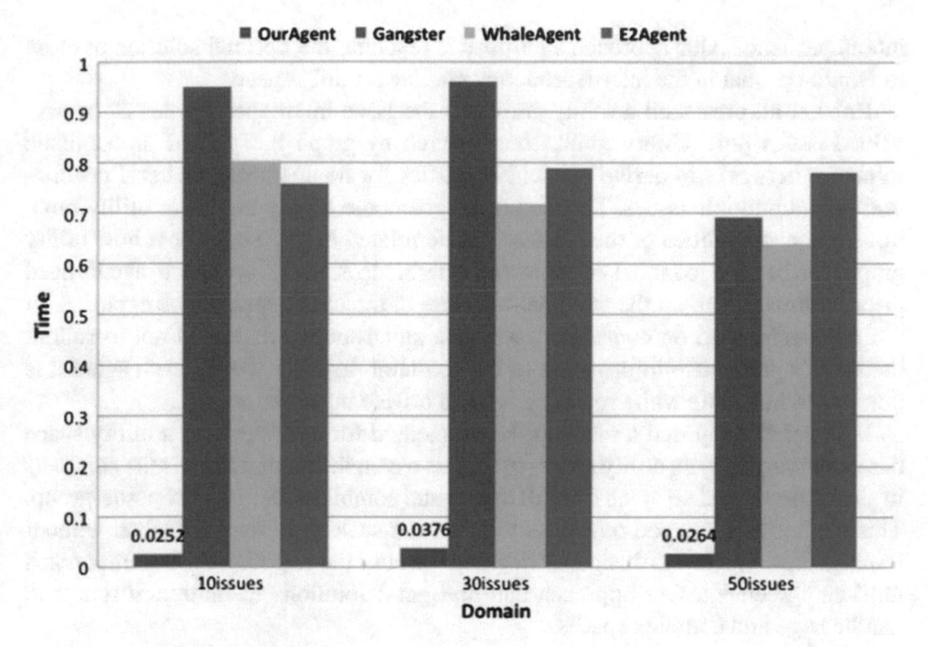

Fig. 6 Average of accept time

5 Related Work

Even though negotiation seems to involve a straightforward distributed constraint optimization problem [11, 23], we have been unable to exploit existing work on high efficiency constraint optimizers. Such solvers attempt to find the solutions that maximize the weights of the satisfied constraints, but do not account for the fact that the final solution must satisfy at least one constraint from every agent.

Fatima et al. proposed an agenda-based model for multi-issue negotiation under time constraints in an incomplete information setting [6]. This paper indicated game theoretic models for issue-by-issue negotiation have two main shortcomings. Firstly, they study the strategic behavior of agents by considering the information they have as *common knowledge*. However, the information that a player has about its opponent is mostly acquired through learning from previous negotiation in fact. Secondly, these models do not consider agent deadlines. Therefore, they need to overcome these problems by considering each agent to have its own deadline and by examining each agent's information state as its *private knowledge*. In this paper proposed agenda-based model, the order in which issues are bargained over and agreements are reached is determined endogenously, as part of the bargaining equilibrium, and each agent's information is its private knowledge. As a result, this paper conducted the properties of the equilibrium solution under which it is unique, symmetric, and Pareto-optimal. However, this paper do not consider about nonlinear utility space which having interdependence. Our approach's purpose is reaching the optimal solution at close to Pareto-optimal in the interdependency nonlinear utility space.

Robu et al. presented a utility graph for the issue interdependencies of binary-valued issues [26]. Utility graphs are inspired by graph theory and probabilistic influence networks to derive efficient heuristics for nonmediated bilateral negotiations about multiple issues. The idea is to decompose highly nonlinear utility functions in the subutilities of the clusters of interrelated items. They show how utility graphs can be used to model an opponent's preferences. In this approach, agents need prior information about the maximal structure of the utility space to be explored. In our approach based on constraints, we have an advantage that need not to submit the agent's preference information to the mediator. In other words, each agent it is possible to negotiate while retaining its own private information.

Fujita et al. proposed a mediator-based method for decomposing a utility space based on every agent's utility space [9]. In this paper, the mediator finds the contracts in each group based on votes from all agents and combines them in each issue-group. This method allows good outcomes with greater scalability than a method without issue-groups. However, this paper does not discuss the negotiation time that reach until an agreement. Our approach can find good solutions in minimum time and handle large and complex spaces.

6 Conclusions

This paper focuses on research in the area of bilateral multi-issue closed negotiations, which is an important class of real-life negotiations. This paper proposed a novel mediator-based negotiation model that can efficiently mediate issue-by-issue negotiation. Our model adopted a dependency-based hypergraphical utility model, and we could perform the negotiations while keeping privacy on the utility values. We demonstrated that the proposed model results in good outcomes for the Pareto-optimal solution and reducing negotiation time.

In our possible future works, efficiency of the mediator-base model in complex utility spaces which has a lot of issues, also needs to be discussed in more detail. Furthermore, we will improve a strategic model that agent compromise strategies. It is necessary to examine the way of effective negotiation based on existing agent strategies.

References

1. T. Baarslag, K. Hindriks, C. Jonker, Acceptance conditions in automated negotiation, in *Complex Automated Negotiations: Theories, Models, and Software Competitions* (Springer, Berlin, 2013), pp. 95–111
2. T. Baarslag, K.V. Hindriks, Accepting optimally in automated negotiation with incomplete information, in *Proceedings of the 2013 International Conference on Autonomous Agents*

and Multi-agent Systems (International Foundation for Autonomous Agents and Multiagent Systems, 2013), pp. 715–722
3. M. Bac, H. Raff, Issue-by-issue negotiations: the role of information and time preference. Games Econ. Behav. **13**(1), 125–134 (1996)
4. P. Faratin, C. Sierra, N.R. Jennings, Using similarity criteria to make issue trade-offs in automated negotiations. Artif. Intell. **142**(2), 205–237 (2002)
5. S.S. Fatima, M. Wooldridge, N.R. Jennings, Multi-issue negotiation under time constraints. in *Proceedings of the First International Joint Conference on Autonomous Agents and Multiagent Systems: Part 1* (ACM, 2002), pp. 143–150
6. S.S. Fatima, M. Wooldridge, N.R. Jennings, An agenda-based framework for multi-issue negotiation. Artif. Intell. **152**(1), 1–45 (2004)
7. C. Fershtman, The importance of the agenda in bargaining. Games Econ. Behav. **2**(3), 224–238 (1990)
8. K. Fujita, T. Ito, M. Klein, A preliminary result on a representative-based multi-round protocol for multi-issue negotiations, in *Proceedings of the 7th International Joint Conference on Autonomous Agents and Multiagent Systems*, vol. 3 (International Foundation for Autonomous Agents and Multiagent Systems, 2008), pp. 1573–1576
9. K. Fujita, T. Ito, M. Klein, An approach to scalable multi-issue negotiation: Decomposing the contract space, in *Computational Intelligence* (2012)
10. K. Gal, T. Ito, C. Jonker, S. Kraus, K. Hindriks, R. Lin, T. Baarslag, The fourth international automated negotiating agents competition (anac2013) (2013), http://www.itolab.nitech.ac.jp/ANAC2013/
11. R. Greenstadt, J.P. Pearce, M. Tambe, Analysis of privacy loss in distributed constraint optimization. AAAI **6**, 647–653 (2006)
12. R. Hadfi, T. Ito, Modeling complex nonlinear utility spaces using utility hyper-graphs, in *Modeling Decisions for Artificial Intelligence* (Springer, Heidelberg, 2014), pp. 14–25
13. R. Inderst, Multi-issue bargaining with endogenous agenda. Games Econ. Behav. **30**(1), 64–82 (2000)
14. T. Ito, H. Hattori, M. Klein, Multi-issue negotiation protocol for agents: Exploring nonlinear utility spaces, in *Proceedings of the 30th International Joint Conference on Artificial Intelligence (IJCAI2007)*, vol. 7 (2007), pp. 347–1352
15. T. Ito, C. Jonker, S. Kraus, K. Hindriks, K. Fujita, R. Lin, T. Baarslag, The second international automated negotiating agents competition (anac2011) (2011), http://www.anac2011.com/
16. T. Ito, K. Fujita, C. Jonker, K. Hindriks, T. Baarslag, R. Aydogan, The fifth international automated negotiating agents competition (anac2014) (2014), http://www.itolab.nitech.ac.jp/ANAC2014/
17. N.R. Jennings, P. Faratin, A.R. Lomuscio, S. Parsons, M.J. Wooldridge, C. Sierra, Automated negotiation: prospects, methods and challenges. Group Decis. Negot. **10**(2), 199–215 (2001)
18. C. Jonker, S. Kraus, K. Hindriks, R. Lin, The first automated negotiating agents competition (anac2010) (2010), http://mmi.tudelft.nl/negotiation/tournament
19. S. Kawaguchi, K. Fujita, T. Ito, Compromising strategy based on estimated maximum utility for automated negotiation agents competition (anac-10). in *Modern Approaches in Applied Intelligence* (Springer, Heidelberg, 2011), pp. 501–510
20. M. Klein, P. Faratin, H. Sayama, Y. Bar-Yam, Negotiating complex contracts. Group Decis. Negot. **12**(2), 111–125 (2003)
21. S. Kraus, *Strategic Negotiation in Multiagent Environments* (MIT press, Cambridge, 2001)
22. S. Kraus, J. Wilkenfeld, G. Zlotkin, Multiagent negotiation under time constraints. Artif. Intell. **75**(2), 297–345 (1995)
23. R.T. Maheswaran, J.P. Pearce, P. Varakantham, E. Bowring, M. Tambe, Valuations of possible states (vps): a quantitative framework for analysis of privacy loss among collaborative personal assistant agents, in *Proceedings of the Fourth International Joint Conference on Autonomous Agents and Multiagent Systems* (ACM, 2005), pp. 1030–1037
24. M.J. Osborne, A. Rubinstein, *Bargaining and Markets*, vol. 34 (Academic press, San Diego, 1990)

25. M.J. Osborne, A. Rubinstein, *A Course in Game Theory* (MIT press, Cambridge, 1994)
26. V. Robu, D. Somefun, J.A. La Poutré, Modeling complex multi-issue negotiations using utility graphs, in *Proceedings of the Fourth International Joint Conference on Autonomous Agents and Multiagent Systems* (ACM, 2005), pp. 280–287
27. A. Rubinstein, Perfect equilibrium in a bargaining model. Econom.: J. Econom. Soc. **50**, 97–109 (1982)
28. A. Rubinstein, A bargaining model with incomplete information about time preferences. Econom.: J. Econom. Soc. **53**, 1151–1172 (1985)
29. T. Sandholm, N. Vulkan, et al, Bargaining with deadlines. in *AAAI/IAAI* (1999), pp. 44–51
30. C.R. Williams, V. Robu, E.H. Gerding, N.R. Jennings, Using gaussian processes to optimise concession in complex negotiations against unknown opponents. in *IJCAI Proceedings-International Joint Conference on Artificial Intelligence*, vol. 22 (2011), pp. 432–438
31. C.R. Williams, V. Robu, E. Gerding, N.R. Jennings, T. Ito, C. Jonker, S. Kraus, K. Hindriks, R. Lin, T. Baarslag, The third international automated negotiating agents competition (anac2012) (2012), http://anac2012.ecs.soton.ac.uk/

Using Graph Properties and Clustering Techniques to Select Division Mechanisms for Scalable Negotiations

Ivan Marsa-Maestre, Catholijn M. Jonker, Mark Klein and Enrique de la Hoz

Abstract This paper focuses on enabling the use of negotiation for complex system optimisation, which main challenge nowadays is scalability. Our hypothesis is that analysing the underlying network structure of these systems can help divide the problems in subproblems which facilitate distributed decision making through negotiation in these domains. In this paper, we verify this hypothesis with an extensive set of scenarios for a proof-of-concept problem. After selecting a set of network metrics for analysis, we cluster the scenarios according to these metrics and evaluate a set of mediation mechanisms in each cluster. The validation experiments show that the relative performance of the different mediation mechanisms change for each cluster, which confirms that network-based metrics may be useful for mechanism selection in complex networks.

1 Introduction

A wide range of real world systems can be modelled as dynamic sets of interconnected nodes [13, 21]. The adequate management of complex networked systems is becoming critical for industrialized countries, since they keep growing in size and complexity. An important sub-class involves autonomous, self-interested enti-

I. Marsa-Maestre (✉) · E. de la Hoz
University of Alcala, Madrid, Spain
e-mail: ivan.marsa@uah.es

E. de la Hoz
e-mail: enrique.delahoz@uah.es

C.M. Jonker
Technical University of Delft, Delft, The Netherlands
e-mail: c.m.jonker@tudelft.nl

M. Klein
Massachusetts Institute of Technology, Cambridge, MA, USA
e-mail: m_klein@mit.edu

© Springer International Publishing AG 2017

K. Fujita et al. (eds.), *Modern Approaches to Agent-based Complex Automated Negotiation*, Studies in Computational Intelligence 674,
DOI 10.1007/978-3-319-51563-2_5

ties (e.g. drivers in a transportation network). The self-interested nature of the entities in the network causes the network to deviate from socially-optimal behaviour. This leads to problems related to unavailability and inefficient use of resources, such as severe traffic jams or casualties in evacuations. New techniques are needed to manage these exponentially growing complex self-interested networks (CSIN) that form the social infrastructures we rely on for progress and welfare. Different fields of research are working on these challenges, but, so far, with only mixed success. Optimization techniques are especially suited to address large-scale systems with an underlying network structure, usually with a divide and conquer approach [27, 32]. However, their performance severely decreases as the complexity of the system increases [23], and with the presence of autonomous entities which deviate from the globally optimal solution, thus harming the social goal. Negotiation techniques are known to be useful to handle self-interested behaviour, but scale poorly with problem size and the intricacies of interdependencies [14]. We focus on distributed, mediated solutions, where a mediator first divides the problem into interconnected subproblems, and then the agents interact (by means of negotiation techniques) to evolve into a solution by themselves. However, given the wide variety of CSIN domains and the inherent variability of CSIN scenarios even within a single domain, intending to find a one-size-fits-all mechanism is unrealistic. Instead, our hypothesis is that the underlying network structure may be used to characterize CSIN scenarios, and to select the most adequate mechanism for each scenario. In this paper, we contribute to test this hypothesis in the following way:

- We propose a proof-of-concept domain for CSINs ("chessboard-evacuation"), and generate a number of scenarios in different categories for it (Sect. 3).
- We select a set of metrics based on graph theory to analyze these scenarios (Sect. 4).
- We cluster the scenarios according to the aforementioned metrics, and then apply a collection of distributed, mediated division approaches to each cluster. Experiments show how the relative performance of the different mediation mechanisms change for each cluster (Sect. 5).

2 Complex Self-interested Networks (CSIN)

Network models are a suitable way to represent many real-world systems [22, 24]. This paper focuses on a particular set of networked systems, where network structure and element behaviour may change dynamically, where there is a social goal or desired behaviour for the network as a whole, and where there are autonomous elements (agents) with individual objectives (also called preferences or utility spaces), usually in conflict with the social goal or among themselves. There are a great number of real-world problems fitting into this category, like electricity grids, transportation infrastructures or cellular communication systems. We call these systems Complex Self-Interested Networks (CSIN). The problem of achieving efficient behaviours in these systems in terms of both social and individual goals is what we call CSIN behaviour optimization. Techniques potentially suited for CSIN include auctions, optimization techniques, and negotiation protocols. Combinatorial auctions [26, 35] can enable large-scale collective decision-making in nonlinear domains, but only of limited type (i.e., negotiations consisting solely of resource/item allocation decisions). Multi-attribute auctions [1, 4, 12] are also aimed only at purchase negotiations and require full revelation of preference information. Constraint-based and other optimization tools [3, 18, 33] offer good solutions with interdependent issues, but are not equipped to deal with self-interested parties. The distributed, adaptive, and self-interested nature of CSIN suggests the use of negotiation techniques. The negotiation research literature offers solutions for problems with one issue (typically price) or a few independent issues [1, 7, 9, 25]. However, these solutions are demonstrably sub-optimal for negotiations with multiple interdependent issues [14]. Attempts to address this challenge [11, 16, 36] face serious limitations in terms of outcome optimality, strategic stability and scalability. These three criteria are key performance indicators for the success of optimization systems in real-world CSIN infrastructures, due to their continuous increase in network size, structural complexity and dynamics [28]. Promising results in overcoming the aforementioned limitations were achieved by developing negotiation mechanisms suited for complex system optimization. These negotiation mechanisms build on techniques from computer science (e.g. nonlinear optimization, complexity analysis), which enable a clever search of the agent utility spaces. This allows to reduce the combinatorial size of the problem and to increase the optimality of the negotiation outcomes (agreements). However, due to the high diversity of complex negotiation scenarios, the different approaches are specifically tailored for particular domains, and it is very unlikely to find an approach which can be used to tackle any arbitrary complex system [17]. However, real world CSIN problems are not arbitrary; they have an underlying network structure. Our hypothesis is that this structure can be exploited to select the most adequate mediation mechanisms from a library of available approaches.

3 Proof-of-Concept Domain: Chessboard Evacuation

To test our hypothesis, we have devised a proof-of-concept domain for a preliminary validation experiment. This is what we have called the chessboard evacuation problem, which has some resemblances with the coordinating pebble motion on graphs problem [5, 15], and with cooperative robot path finding [29]. An example chessboard evacuation scenario is shown in Fig. 1.

Let us assume that pawns want to evacuate the chessboard as quickly as possible through the exit represented with an arrow. Pawns can move one square per time unit in the vertical or horizontal axis, and they lose time when they collide. This problem has the main characteristics of CSIN:

- The coordinated pebble motion and the cooperative robot path finding problems are known to be NP-hard [10, 29].

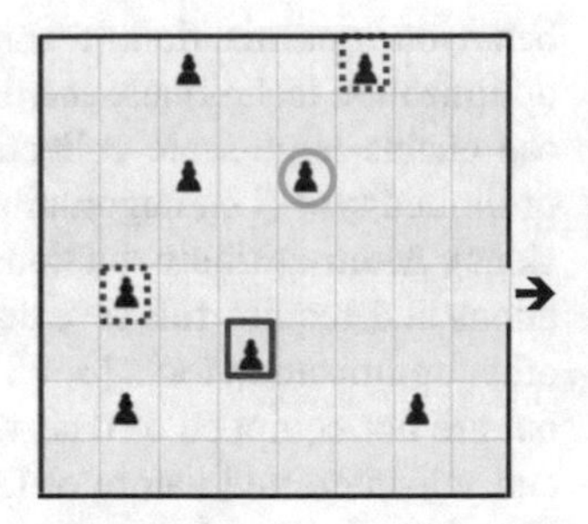

Fig. 1 Chessboard evacuation scenario

- Once a solution has been found, some pawns may disagree on having to wait or to take longer than optimal paths to avoid conflicts, which may make them deviate from the solution, causing collisions and inefficient behaviours.

3.1 Formalisation of the Problem

An instance of the chessboard evacuation problem can be seen as a tuple $\langle B, P, g\rangle$, where:

- $B = \langle N, E\rangle$ is a graph representing the board, where each node $n \in N$ is a space in the board, and each edge $e \in E$ connects two adjacent spaces. Furthermore, for any two nodes $n, m \in N$, $e(n, m)$ denotes an edge between n and m.
- P is the set of pawns. Each pawn p is characterized by its initial position $n_{p,0} \in N$, which refers to a node in the graph B.
- $g \in N$ is the goal, representing the square from which pawns will evacuate the board.

A potential solution to the problem would be a set of routes $R = \{r_p | p \in P\}$, where each route $r_p = \{n_{p,t} \in N | t = 0, \ldots, \tau_p\}$ represents the sequence of positions occupied by pawn p at each time t. The evacuation time for pawn p is denoted by τ_p. For a solution R to be valid, all pawns need to travel a continuous path from their initial position to the goal and no pawns can be allowed to be at the same position at the same time, that is:

- $n_{p,\tau_p} = g \forall p$
- Consecutive positions in a route correspond either to the same node or to connected nodes in the graph, that is, nodes which are connected by an edge:

$$\forall t, p : e(n_{p,t}, n_{p,t+1}) \in E \text{ or } n_{p,t} = n_{p,t+1} \quad (1)$$

- $\forall t, \forall p, p' \in P : r_{p,t} \neq r_{p',t}$

For any solution R, the time $\tau(R) = \max_{p \in P} |\tau_p|$ is called evacuation time. A solution R_i is assumed to be better than another solution R_j if $\tau(R_i) < \tau(R_j)$.

3.2 Modeling Agent Self-interests

Of course, in complex self-interested networks, solutions are not implemented either in a centralized or in a one-shot manner. Even if solutions are computed in a centralized way, agents are self-interested and autonomous, and can deviate at any point from any externally imposed plan. To represent this in the chessboard problem, we model agent decision making as follows:

- Pawns give a value v to each node $n \in N$, equal to the shortest path length from this node to the goal g. Therefore, pawns self-interest is normally to move to the lowest value adjacent node.
- Pawns are considered to be in conflict if the values of the nodes corresponding to their current positions are equal, since that means those pawns would necessarily collide in their way to the goal if they take their optimal paths and do not wait.
- At any time t, pawns which are asked (by imposition or negotiation) to make a move which is suboptimal (i.e., does not minimize value of the next node), will follow their optimal path (i.e., they will deviate from the proposed solution) with probability

$$\pi = 1 - \frac{1}{n_c}, \tag{2}$$

  where n_c is the expected number of conflicts the pawn would be in in the next time unit if it followed the suggested route. This models pawn risk aversion, in the sense that being in a high number of conflicts increases the expected evacuation time. Note that, if the proposed route does not involve conflicts for the pawn, it will follow it without any problem.

Since pawns are still self-interested as defined above, they may produce collisions. If two or more pawns collide in the same node, they are all sent back to the node they came from. Collisions propagate backwards, so any pawns trying to enter a node where another pawn has been sent back due to a collision would also suffer a collision.

3.3 Categories of Scenarios

For the verification of the hypothesis to be significant enough, it has to be evaluated in a wide variety of scenarios. In order to achieve this variety, we have considered three different categories of scenarios, according to the properties of the underlying graph B:

- *Chessboards (CB)*: in this category, the graph is a 8-by-8 square lattice (resembling a chessboard), where we randomly add a number of obstacles (non-transitable nodes) for diversity.

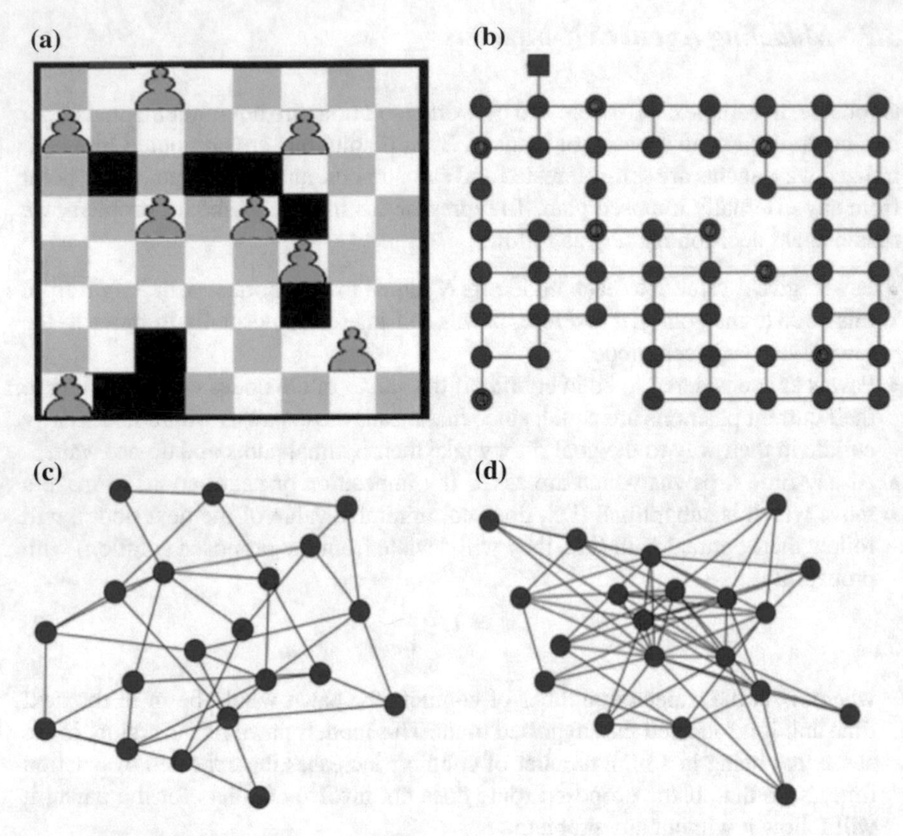

Fig. 2 Examples of each generated graph category: **a** chessboard with 8 obstacles; **b** corresponding lattice graph; **c** ER graph with $p = 0.2$; **d** BA graph with $\alpha = 4$

- *Erdös–Renyi graphs (ER)*: in this category, graphs are generated randomly using the Erdös–Renyi model [20]. In the ER model, every pair of nodes $\{m, n \in N\}$ has a finite probability p of being connected.
- *Barabasi–Albert graphs (BA)*: in this category, graphs are generated randomly using the Barabasi–Albert model [20]. In the BA model, after an initial number of nodes $|N|_0$ has been placed, subsequent nodes are added one-by-one to the graph, with each new node being connected to exactly α of the existing nodes. The probability of a new node connecting to an existing node i is $p_i = \frac{k_i}{\sum_j k_j}$, where k_i is the degree of existing node i. This is called preferential attachment.

For each category of scenarios, different sub-categories were created by varying the number of obstacles, the probability of connection p, and the attachment parameter α. In each scenario, a node is randomly chosen as the goal (in the case of chessboards, one of the peripheral nodes). Figure 2 shows an example for each category of graphs.

4 Graph Metrics for Scenario Characterization

With the model for self-interested agents described above, we run a number of simulations in the aforementioned scenario categories. The details of these experiments are beyond the scope of this paper, and can be found in [31]. The experiments allowed us to select a number of graph metrics from the literature which were significantly correlated to the evacuation times of the simulations. The selected metrics are the following:

- *Graph order*: the number of nodes in the graph.
- *Graph diameter*: the longest distance between any pair of nodes in the graph [19].
- *Wiener index*: gives a measure of graph complexity from the distances in the graph. It is computed as $W(G) = \frac{1}{2}\sum_{i=0}^{|N|}\sum_{j=0}^{|N|} d(n_i, n_j)$, where $d(n_i, n_j)$ is the shortest distance between nodes n_i and n_j [34].
- *Graph density*: the ratio between the number of edges in the graph and the maximum number of edges (that is, if it were a fully-connected graph). For undirected graphs, this density is computed as $D = \frac{2|E|}{|N|(|N|-1)}$.
- *Clustering coefficient*: a measure of the degree to which nodes in a graph tend to cluster together. The cluster coefficient of a graph is computed as the average of the local clustering coefficient of its nodes, which is computed for each node as the ratio between the number of links between its neighbors and the maximum number of links between them (that is, if they were fully connected).
- *Assortativity*: the correlation between the degree (number of neighbors) of adjacent nodes [30].
- *Entropy of betweenness centrality*: Centrality metrics measure the importance of a node within a graph. In particular, betweenness centrality of a node is the ratio of shortest paths in the graph which traverse the node. It is computed as $C_B(n) = \sum_{s,t \in V} \frac{\sigma(s,t|n)}{\sigma(s,t)}$, where $\sigma(s, t)$ is the number of shortest paths between nodes s and t, and $\sigma(s, t|n)$ is the number of such paths where n acts as a bridge. From this metric, we can assess the complexity of a graph using Shannon information theory, by transforming the betweenness centrality metric into a probability function $p(n_i) = \frac{C_B(n_i)}{\sum_{j=0}^{|N|} C_B(n_j)}$, and computing the entropy as $H(G) = -\sum_{i=1}^{|N|} p(n_i)\log(p(n_i))$.

These metrics have been used to divide the scenarios in clusters for the experiments, as described in the following section.

5 Using Graph Metrics for Mechanism Selection

Our hypothesis is that the selected set of metrics may be used as a basis for mechanism selection in CSIN problems, and in particular in the evacuation problem considered in this paper. To verify this, we have devised a number of distributed mediation mechanisms, and we have performed an extensive set of experiments with them in the generated scenarios. We have also clustered the scenarios according to the aforementioned metrics, so that we can see the influence of these metrics in the performance of the different mechanisms.

5.1 Distributed, Mediated Division Approaches

As mentioned above, our interest regarding mechanisms focuses on distributed, mediated solutions, where a mediator first divides the problem into interconnected subproblems, and then the agents are let to evolve into a solution by themselves. The two key factors here are, first, how to divide the problem, and then, how to interconnect the different subproblems so that the self-interested agents can autonomously evolve to an emergent, efficient solution. In the particular case of the chessboard evacuation problem, we have chosen to dynamically divide the chessboard graph into subgraphs (lets call them worlds), and let the pawns in each world negotiate the paths they will take. In addition, we make pawns within a world negotiate about where to place the entrance arrows to their world, as seen in Fig. 3a. This entrance arrow placement is the negotiation technique (very simple, in this case) which interconnects the subproblems and guarantees emergence and incentive for cooperation. Since pawns can govern how other pawns enter their worlds, they can ensure that the pawns conceding in a negotiation (i.e., sacrificing their own utility to solve a conflict) are not exposing themselves to further conflicts. In this way, the number of conflicts in a world never increases. This invariant guarantees the progress of the approach, in contrast with the situation we had with the exponential increase due to collisions. An example of this is depicted in Fig. 3b, where pawns within the lower-right world (LR) are in conflict. Any pawn which concedes to the other would be then in conflict with the pawn in the lower-left world (LL). However, they have placed the entrance from LL to LR ensuring that this conflict is no longer possible, since any pawn coming from LL would be further from the goal than any pawn in LR. Therefore, a pawn in LR can agree to wait with the guarantee that its utility loss is bounded (one time unit). This provides an incentive both to accept the proposed problem division and to cooperate within it.

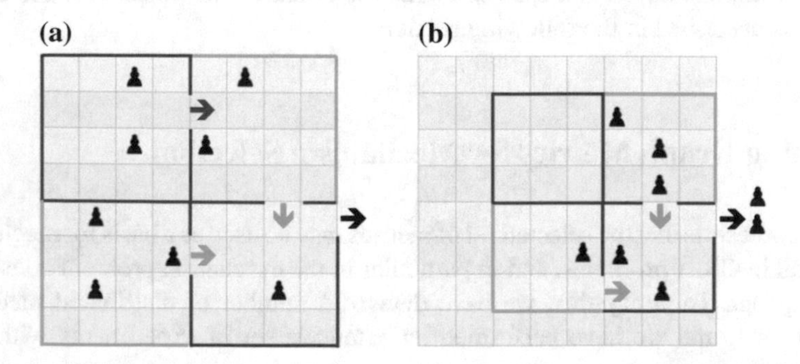

Fig. 3 Distributed, mediated division example

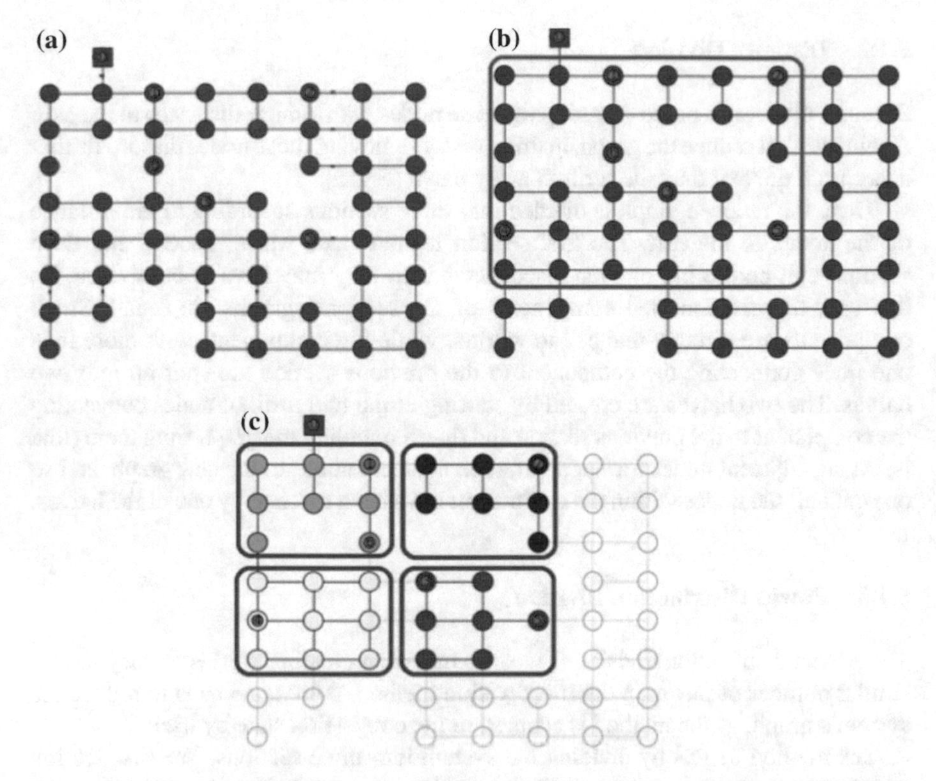

Fig. 4 Example of symmetric division. **a** Chessboard to divide. **b** Minimum square containing all the pawns and the exit. **c** Worlds generated

5.1.1 Symmetric Division

The symmetric division (Fig. 4) is based on dividing the system into four symmetric subsystems. To apply this division method to the chessboard scenarios, first the graph is reduced by getting the minimum square that contains all the pawns and the exit of the system. By using this graph reduction, the evacuation plan is adapted to the congestion state at any given moment. Once the graph has been reduced, the reduced graph is symmetrically divided into four symmetric subgraphs, creating four worlds.

The position of the random graphs' nodes is not fixed, so there can be many different representations of a network. Thus, to apply the symmetric division to random graphs, we need to assign a position on the plane for each node. For that purpose, the Fruchterman–Reingold algorithm is used [8].

5.1.2 Distance Division

Distance division is based on grouping close nodes with similar distances to the exit. Again, we first reduce the graph, in this case by removing those nodes that are further away from the exit than the furthest away pawn (Fig. 5).

Then, the reduced graph is divided into three sections according to the distance of the nodes to the exit. The first section leads to one world. Second and third sections can be too big or even disconnected, so they may have to be divided. To this end, the disconnected components of the sections' graphs are found. Small components are directly mapped to worlds, while big components with more than one node connecting the component to the previous section are split up into two halves. The two halves are created by starting at the two furthest nodes connecting the component to the previous section and then expanding the graph from them (that is, taking adjacent nodes further apart, then adjacent nodes to the new graph, and so on) until all the nodes within the component have been reached by one of the halves.

5.1.3 Pawns Distribution Division

The pawns distribution division (Fig. 6) is based on creating worlds so they have a similar number of pawns. With this division method, there is no need to reduce the system's graph, as the method is adapted to the congestion state by itself.

The method begins by dividing the system into three sections. The first section is created by starting at the exit and expanding the graph from it until 1/5 of the pawns are reached. The second section is generated starting at the nodes connecting the section to the previous section and expanding the graph from them until 2/5 of the pawns are reached. The third section contains the remaining nodes. Once the sections have been generated, the first section is mapped to a world, while the second and third sections are divided. As in the previous method, first the disconnected components of the sections are found. Those components with less than half the number of pawns of the corresponding section are directly mapped to worlds, while components with more pawns and more than one node connecting the component to the previous section are divided into two worlds, each one containing half the pawns of the component.

5.1.4 Community Division

In graph theory, a community is a set of nodes such that there are many connections between nodes of the same community and few connections with nodes outside the community [19]. The community division consist on reducing the graph by removing those nodes further from the exit than the furthest pawns, and, afterwards, using the Louvain algorithm [2] to find the graph communities, map each of these communities to a world (Fig. 7).

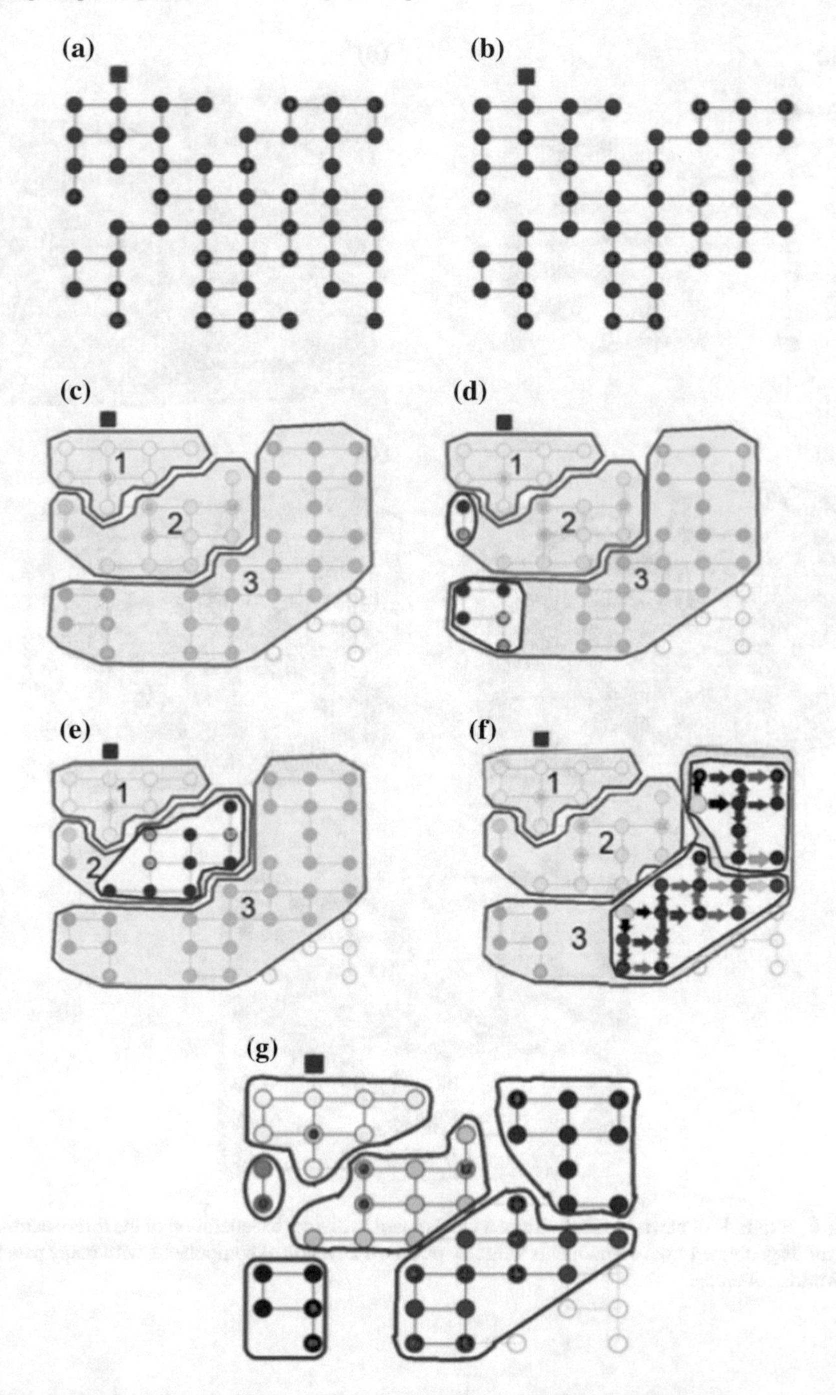

Fig. 5 Example of distance division. **a** Chessboard to divide. **b** Reduced graph. **c** Sections generated. **d** Worlds created from small components. **e** Wolds created from components with just one node connecting the component with the previous section. **f** Division of big components with more than one node connection the component with the previous section. **g** Worlds generated

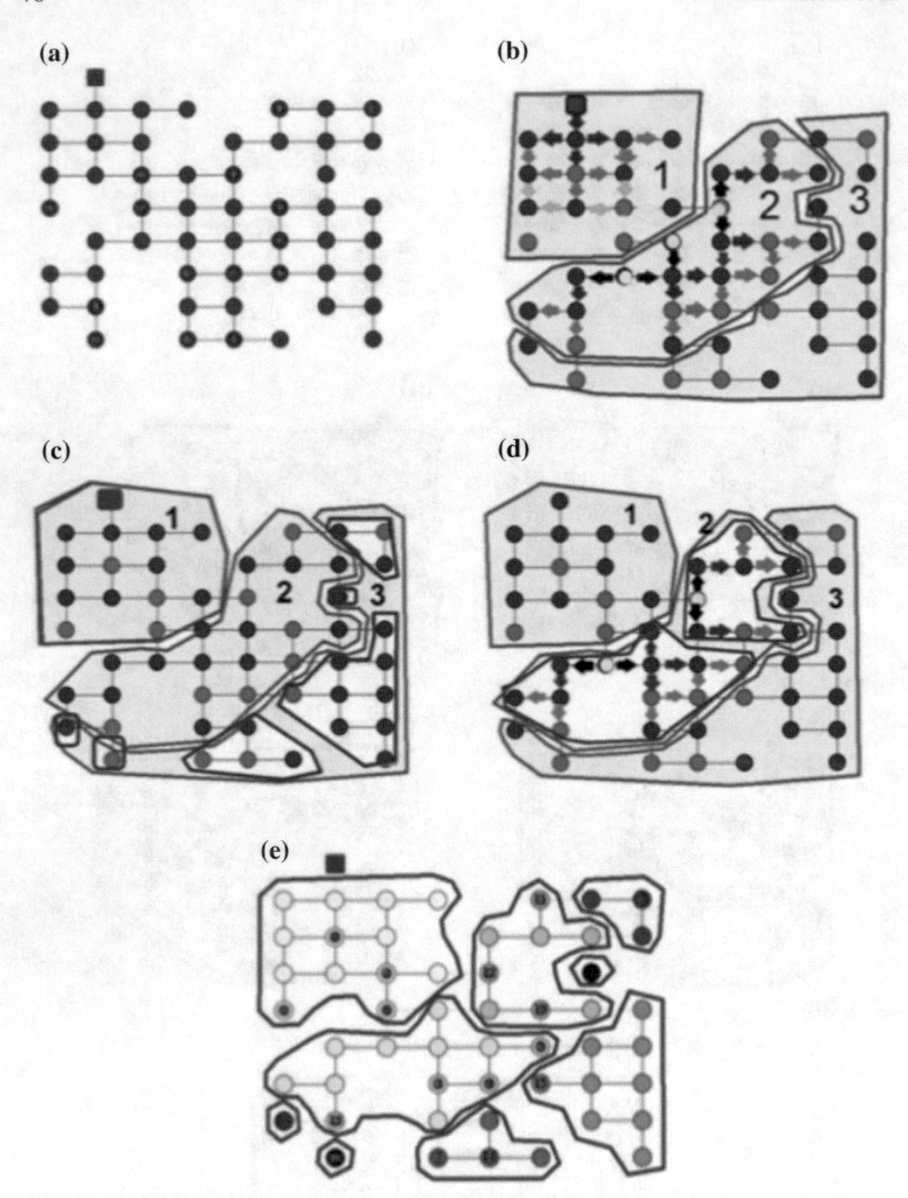

Fig. 6 Example of pawn-based division. **a** Chessboard to divide. **b** Generation of the three sections. **c** Worlds generated from components with few pawns. **d** Division of components with many pawns. **e** Worlds generated

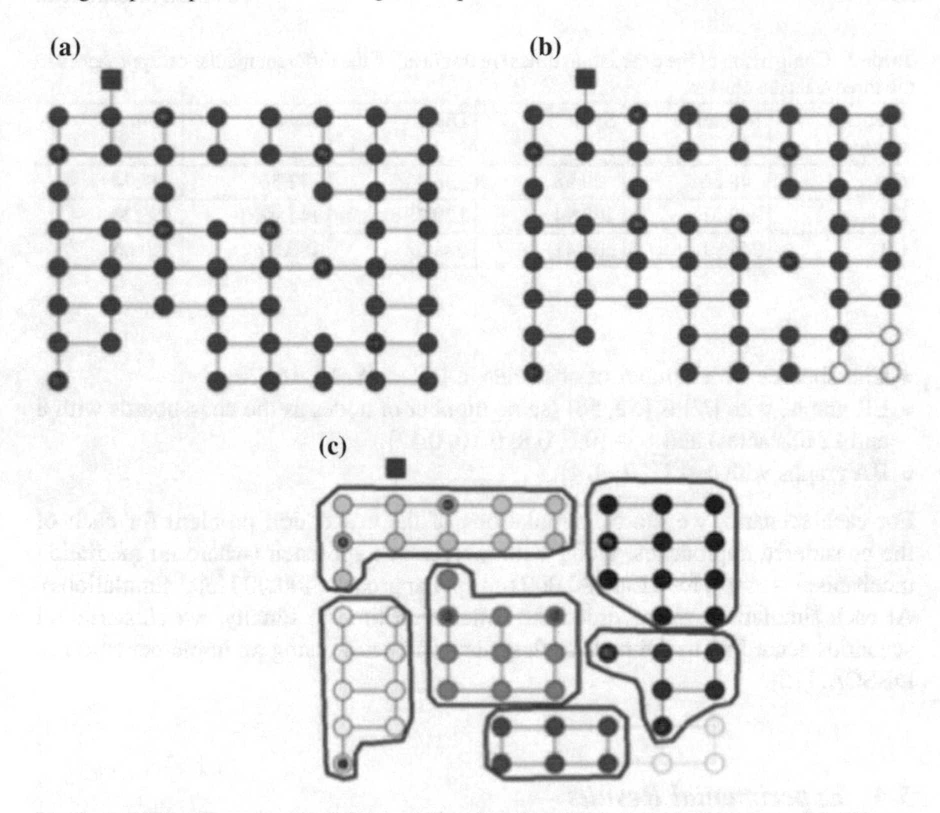

Fig. 7 Example of community division. **a** Chessboard to divide. **b** Reduced graph. **c** Worlds generated

5.2 Negotiation Between Agents

Once the division has been made, agents need to negotiate about where to placc the doors between worlds. Since each possible door-placing schema is a binary vector (with one bit per boundary square, which can be set to 1 or 0 depending on whether there should be a door or not in it, respectively), we have used the approach in [14], which is a negotiation mechanism specifically designed for negotiation of complex contracts with binary issues.

5.3 Experimental Setting

We have generated a total of 1600 scenarios for the categories described in Sect. 3. In particular, we generated 50 scenarios for each combination of the following parameters per category:

Table 1 Comparison of the evacuation times (in minutes) of the different mediation approaches in the three detected clusters

Scenario cluster	Reference	Sym.	Dist.	Pawn	Com.
C1	48.56	29.98	36.63	42.56	**27.44**
C2	240.51	163.94	**129.75**	144.92	156.20
C3	264.91	206.42	288.62	**180.60**	220.66

- Chessboards with number of obstacles in $|O| = \{8, 12, 16, 20\}$.
- ER graphs with $|N| \in \{52, 56\}$ (same number of nodes as the chessboards with 8 and 12 obstacles) and $p \in \{0.6, 0.8, 0.10, 0.12\}$.
- BA graphs with $\alpha \in \{1, 2, 3, 4\}$.

For each scenario, we ran 50 simulations of the evacuation problem for each of the considered approaches, along with a reference approach (where no mediation mechanism is used), for a total 80,000 runs per approach (400,000 total simulations). At each simulation, we recorded the evacuation time τ. Finally, we clustered all scenarios according to the metrics described in Sect. 4, using an implementation of DBSCAN [6].

5.4 Experimental Results

The DBSCAN algorithm yielded three scenario clusters (C1 to C3). Table 1 shows the evacuation times for the different mediation approaches, averaged for each of the scenario clusters. We have also represented the average times for the reference approach. In general, we can see that all mediation approaches outperform the reference unmediated approach (except for a slight disadvantage in using the distance-based approach in the C3 cluster). However, we can see there are significant differences in the relative performance of the approaches for each cluster. In the C1 cluster, symmetric and community based divisions work better than the other approaches (above 20% difference). In the C2 cluster, distance-based division improves over 11% with respect to the next top approach, and in the C3 cluster, pawn-based division outperforms the other approaches by 12.6%. We can conclude that graph metrics can help to select the appropriate mechanism to use when facing a given scenario, providing a significant advantage on average evacuation times than choosing a different strategy or no strategy at all.

6 Discussion and Conclusions

The main hypothesis of our work is that underlying structural properties in Complex Self-Interested Networks (CSIN) can be used to decide which mechanisms to use to enhance the performance of such networks. To validate this hypothesis, in this paper we present a chessboard evacuation problem as a proof of concept domain for CSINs. For this domain, we systematically generate a wide variety of scenarios in three categories of graph structures (lattices, Erdös–Renyi graphs and Barabasi–Albert graphs), and characterize them according to a set of metrics selected from graph theory. Then we consider a number of mediated, distributed approaches to facilitate reaching efficient solutions to the evaluation problems, and run an extensive set of simulations with all of them. We analyse the simulation data by clustering the scenarios using the aforementioned metrics. Experiments show that the relative performance of the

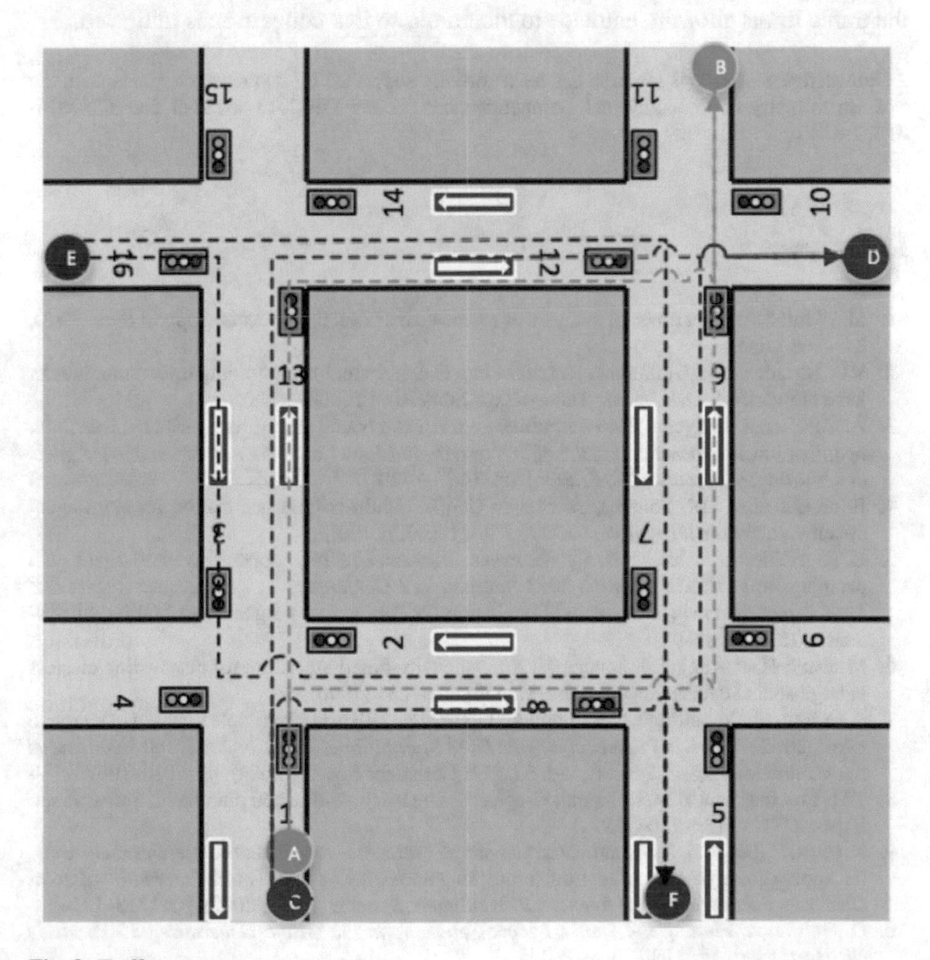

Fig. 8 Traffic management scenario

different mediation approaches significantly change in each cluster, which backs up the idea that analysing the graph properties of a problem can help choosing a suitable mechanism to address it. Though our experiments yield promising results, there is still plenty of work to be done in this area. We are in the process to validate the suitability of the metrics to perform a priori mechanism selection, by training classifiers (e.g. random trees) in large sets of scenarios and using them to predict which would be the best approach to use when confronted with new scenarios. We also want to get more diverse scenario sets, since in this case the clusters detected by DBSCAN where populated in their majority (about 90%) by scenarios in a single generation category (either chessboards, Erdös–Renyi or Barabasi–Albert graphs), which demonstrates a significant generation bias in the experiment set. Finally, we are interested in generalizing the approach to other domains out of the chessboard proof-of-concept. For instance, we are working on a transportation management scenario (Fig. 8), where GPS navigator agents within a given area automatically negotiate about the state of the traffic lights allowing entrance to their area, so that congestion is mitigated.

Acknowledgements This research has been partially supported by two research grants from the Spanish Ministry of Economy and Competitiveness (grants TEC2013-45183-R and TIN2014-61627-EXP).

References

1. M. Bichler, An experimental analysis of multi-attribute auctions. Decis. Support Syst. **29**(3), 249–268 (2000)
2. V.D. Blondel, J.-L. Guillaume, R. Lambiotte, E. Lefebvre, Fast unfolding of communities in large networks. J. Stat. Mech.: Theory Exp. **2008**(10), P10008 (2008)
3. A. Chechetka, K. Sycara, No-commitment branch and bound search for distributed constraint optimization, in *Proceedings of the Fifth International Joint Conference on Autonomous Agents and Multiagent Systems* (ACM, 2006), pp. 1427–1429
4. E. de Oliveira, J.M. Fonseca, A. Steiger-Garção, Multi-criteria negotiation on multi-agent systems, in *Proceedings of the CEEMAS'99* (1999), p. 190
5. B. de Wilde, A.W. ter Mors, C. Witteveen, Push and rotate: cooperative multi-agent path planning, in *Proceedings of the 2013 International Conference on Autonomous Agents and Multi-agent Systems* (International Foundation for Autonomous Agents and Multiagent Systems, 2013), pp. 87–94
6. M. Ester, H.-P. Kriegel, J. Sander, X. Xu, A density-based algorithm for discovering clusters in large spatial databases with noise. Kdd **96**, 226–231 (1996)
7. S. Fatima, M. Wooldridge, N.R. Jennings, Optimal negotiation of multiple issues in incomplete information settings, in *Proceedings of the Third International Joint Conference on Autonomous Agents and Multiagent Systems*, vol. 3 (IEEE Computer Society, 2004), pp. 1080–1087
8. T.M. Fruchterman, E.M. Reingold, Graph drawing by force-directed placement. Softw. Pract. Exper. **21**(11):1129–1164 (1991)
9. N. Gatti, F. Amigoni, A decentralized bargaining protocol on dependent continuous multi-issue for approximate pareto optimal outcomes, in *Proceedings of the Fourth International Joint Conference on Autonomous Agents and Multiagent Systems* (ACM, 2005), pp. 1213–1214
10. O. Goldreich, *Finding the Shortest Move-Sequence in the Graph-Generalized 15-Puzzle is NP-Hard* (Springer, Heidelberg, 2011)

11. K. Hindriks, C. Jonker, D. Tykhonov, Avoiding approximation errors in multi-issue negotiation with issue dependencies, in *Proceedings of the 1st International Workshop on Agent-based Complex Automated Negotiations (ACAN 2008)* (Citeseer, 2008), pp. 1347–1352
12. J. Kalagnanam, D.C. Parkes, Auctions, bidding and exchange design, in *Handbook of Quantitative Supply Chain Analysis* (Springer, New York, 2004), pp. 143–212.
13. R. Kinney, P. Crucitti, R. Albert, V. Latora, Modeling cascading failures in the North American power grid. Eur. Phys. J. B-Condens. Matter Complex Syst. **46**(1), 101–107 (2005)
14. M. Klein, P. Faratin, H. Sayama, Y. Bar-Yam, Negotiating complex contracts. Group Decis. Negot. **12**(2), 111–125 (2003)
15. D. Kornhauser, G. Miller, P. Spirakis, in *Coordinating Pebble Motion on Graphs, the Diameter of Permutation Groups, and Applications* (IEEE, 1984)
16. M. Li, Q.B. Vo, R. Kowalczyk, Searching for fair joint gains in agent-based negotiation, in *Proceedings of the 8th International Conference on Autonomous Agents and Multiagent Systems*, vol. 2 (International Foundation for Autonomous Agents and Multiagent Systems, 2009), pp. 1049–1056
17. I. Marsa-Maestre, M. Klein, C.M. Jonker, R. Aydoğan, From problems to protocols: towards a negotiation handbook. Decis.Support Syst. **60**, 39–54 (2014)
18. P.J. Modi, W.-M. Shen, M. Tambe, M. Yokoo, Adopt: asynchronous distributed constraint optimization with quality guarantees. Artif. Intell. **161**(1), 149–180 (2005)
19. M. Newman, *Networks: An Introduction* (Oxford University Press, Oxford, 2010)
20. M.E. Newman, D.J. Watts, S.H. Strogatz, Random graph models of social networks. Proc. Natl. Acad. Sci. **99**(suppl 1), 2566–2572 (2002)
21. C. Osorio, M. Bierlaire, Mitigating network congestion: analytical models, optimization methods and their applications, in *90th Annual Meeting*, number EPFL-TALK-196049 (2011)
22. G. Palla, I. Derényi, I. Farkas, T. Vicsek, Uncovering the overlapping community structure of complex networks in nature and society. Nature **435**(7043), 814–818 (2005)
23. M. Pelikan, K. Sastry, D.E. Goldberg, Multiobjective estimation of distribution algorithms, in *Scalable Optimization via Probabilistic Modeling*, (Springer, Berlin, 2006), pp. 223–248
24. E. Ravasz, A.-L. Barabási, Hierarchical organization in complex networks. Phys. Rev. E **67**(2), 026112 (2003)
25. F. Ren, M. Zhang, Bilateral single-issue negotiation model considering nonlinear utility and time constraint. Decis. Support Syst. **60**, 29–38 (2014)
26. Y. Sakurai, M. Yokoo, K. Kamei, An efficient approximate algorithm for winner determination in combinatorial auctions, in *Proceedings of the 2nd ACM Conference on Electronic Commerce* (ACM, 2000), pp. 30–37.
27. B. Schaller, New York citys congestion pricing experience and implications for road pricing acceptance in the united states. Transp. Policy **17**(4), 266–273 (2010)
28. S.H. Strogatz, Exploring complex networks. Nature **410**(6825), 268–276 (2001)
29. P. Surynek, An optimization variant of multi-robot path planning is intractable, in *Proceedings of the AAAI* (2010)
30. P. Van Mieghem, H. Wang, X. Ge, S. Tang, F. Kuipers, Influence of assortativity and degree-preserving rewiring on the spectra of networks. Eur. Phys. J. B **76**(4), 643–652 (2010)
31. M. Vega, Aplicacin de teora de grafos a redes con elementos autnomos. Master's thesis, University of Alcala, Spain (2014). (In Spanish)
32. P. Vytelingum, S.D. Ramchurn, T.D. Voice, A. Rogers, N.R. Jennings, Trading agents for the smart electricity grid, in *Proceedings of the 9th International Conference on Autonomous Agents and Multiagent Systems*, vol. 1 (International Foundation for Autonomous Agents and Multiagent Systems, 2010), pp. 897–904
33. H. Wang, S. Liao, L. Liao, Modeling constraint-based negotiating agents. Decis. Support Syst. **33**(2), 201–217 (2002)
34. H. Wiener, Structural determination of paraffin boiling points. J. Am. Chem. Soc. **69**(1), 17–20 (1947)
35. M. Xia, J. Stallaert, A.B. Whinston, Solving the combinatorial double auction problem. Eur. J. Oper. Res. **164**(1), 239–251 (2005)

36. X. Zhang, R.M. Podorozhny, V.R. Lesser, Cooperative, multistep negotiation over a multi-dimensional utility function, *Artificial Intelligence and Soft Computing* (Springer, Berlin, 2000), pp. 136–142

Compromising Strategy Considering Interdependencies of Issues for Multi-issue Closed Nonlinear Negotiations

Shinji Kakimoto and Katsuhide Fujita

Abstract Bilateral multi-issue closed negotiation is an important class of real-life negotiations. In this chapter, we propose an estimating method for the pareto frontier based on the opponent's bids. In the proposed method, the opponent's bid is divided into small elements considering the combinations between issues, and the number of the opponent's proposals is counted to estimate the opponent's utility function. In addition, *SPEA2* is employed in the proposed method to search the pareto optimal bids based on the opponent's estimated utility and the agent's own utility. After that, we propose a compromising strategy for nonlinear utility functions based on the estimating method of the opponent's utility functions. Experimental results demonstrate that our proposed method considering the interdependency between issues can search the pareto optimal bids. In addition, we compare the negotiation efficiency of our proposed agent with ten state-of-the-art negotiation agents that entered the final round of ANAC-2014. Our agent has won by a big margin in the negotiation tournaments because of the estimated opponent's utility and *SPEA2*.

1 Introduction

Negotiation is an important process in forming alliances and reaching trade agreements. Research in the field of negotiation originates in various disciplines including economics, social science, game theory and artificial intelligence (e.g. [6] etc.).

This chapter is based on one of the paper in the proceedings of 7th IEEE International Conference on Service-Oriented Computing and Applications (SOCA 2014) written by Kakimoto and Fujita [11].

S. Kakimoto (✉) · K. Fujita
Institute of Engineering, Tokyo University of Agriculture and Technology,
Fuchu, Japan
e-mail: kakimoto@katfuji.lab.tuat.ac.jp

K. Fujita
e-mail: katfuji@cc.tuat.ac.jp

© Springer International Publishing AG 2017
K. Fujita et al. (eds.), *Modern Approaches to Agent-based Complex Automated Negotiation*, Studies in Computational Intelligence 674,
DOI 10.1007/978-3-319-51563-2_6

Automated agents can be used side-by-side with a human negotiator embarking on an important negotiation task. They can alleviate some of the effort required of people during negotiations and also assist people that are less qualified in the negotiation process. There may even be situations in which automated negotiators can replace the human negotiators. Another possibility is for people to use these agents as a training tool, prior to actually performing the task. Thus, success in developing an automated agent with negotiation capabilities has great advantages and implications.

Motivated by the challenges of bilateral negotiations between automated agents, the automated negotiating agents competition (ANAC) was organized [9]. The purpose of the competition is to facilitate research in the area of bilateral multi-issue closed negotiation. The setup at ANAC is a realistic model including time discounting, closed negotiations, alternative offering protocol, and so on. By analyzing the results of ANAC, the stream of strategies of automated negotiations and important factors for developing the competition have been shown [2]. Also, some effective automated negotiating agents have been proposed through the competitions [3, 5].

A key point in achieving automated negotiation in real life is the non-linearity of the utility functions. Many real-world negotiation problems assume the multiple nonlinear utility function. When an automated negotiation strategy covers the linear function effectively, it is not always possible or desirable in nonlinear situations [14]. In other words, it is still an open and interesting problem to design more efficient automated negotiation strategies against a variety of negotiating opponents in different "nonlinear" negotiation domains.

In this chapter, we propose an estimating method for finding the pareto frontier based on the opponent's bids. In the proposed method, the opponent's bid is divided into small ones considering the combinations between issues, and the number of the opponent's proposals is counted to estimate the opponent's utility function. In addition, *SPEA2* [19], which is based on the genetic algorithm for multiple objective optimization, is employed in our proposed method to search the pareto optimal bids based on the opponent's estimated utility and the agent's own utility. In the experiments, we evaluated the quality of the pareto frontier in our approach measured by the size of the dominant area [18]. The experimental results demonstrate that our proposed method considering the interdependency between issues can search the pareto optimal bids.

We also propose a compromising strategy for nonlinear utility functions based on the estimating method of the opponent's utility functions. The proposed agent has two steps: information-gathering and compromising steps. The agent doesn't accept the opponent's offers and proposes selfish bids to gather the opponent's information for estimating the opponent's utility in the information-gathering step. In the compromising step, the agent proposes the pareto optimal bids estimated by *SPEA2* using the opponent's bids collected by the previous step in order to make agreements. Experimental results demonstrate that our proposed method considering the interdependency between issues can search the pareto optimal bids. In addition, we compare the negotiation efficiency of our proposed agent with ten state-of-the-art negotiation agents that entered the final round of ANAC-2014. Our agent has won by a big margin in the negotiation tournaments because of the estimated opponent's utility and *SPEA2*. In addition, this strategy was implemented as *Random Dance* agent in ANAC-2015 [12].

The remainder of the chapter is organized as follows. First, we describe related works. Second, we show the negotiation environments and nonlinear utility functions. Third, we describe *SPEA2*, which can search the pareto optimal bids effectively, and propose a novel method for estimating the opponent's utility function. Then, we demonstrate the experimental analysis of finding the pareto optimal bids. Finally, we present our conclusions.

2 Related Works

This chapter focuses on research in the area of bilateral multi-issue closed negotiation, which is an important class of real-life negotiations. Closed negotiation means that opponents do not reveal their preferences to each other. Negotiating agents designed using a heuristic approach require extensive evaluation, typically through simulations and empirical analysis, since it is usually impossible to predict precisely how the system and the constituent agents will behave in a wide variety of circumstances. Motivated by the challenges of bilateral negotiations between people and automated agents, the automated negotiating agents competition (ANAC) was organized in 2010 [1]. The purpose of the competition is to facilitate research in the area of bilateral multi-issue closed negotiation.

The declared goals of the competition are (1) to encourage the design of practical negotiation agents that can proficiently negotiate against unknown opponents in a variety of circumstances, (2) to provide a benchmark for objectively evaluating different negotiation strategies, (3) to explore different learning and adaptation strategies and opponent models, (4) to collect state-of-the-art negotiating agents and negotiation scenarios, and make them available to the wider research community. The competition was based on the GENIUS environment, which is a General Environment for Negotiation with Intelligent multi-purpose Usage Simulation [15]. By analyzing the results of ANAC, the stream of strategies of ANAC and important factors for developing the competition have been shown. Baarslag et al. presented an in-depth analysis and the key insights gained from ANAC 2011 [2]. This paper mainly analyzes the different strategies using classifications of agents with respect to their concession behavior against a set of standard benchmark strategies and empirical game theory (EGT) to investigate the robustness of the strategies. It also shows that the most adaptive negotiation strategies, while robust across different opponents, are not necessarily the ones that win the competition. Furthermore, our EGT analysis highlights the importance of considering metrics.

Chen and Weiss proposed a negotiation approach called OMAC, which learns an opponent's strategy in order to predict future utilities of counter-offers by means of discrete wavelet decomposition and cubic smoothing splines [4]. They also presented a negotiation strategy called EMAR for this kind of environment that relies on a combination of Empirical Mode Decomposition (EMD) and Autoregressive

Moving Average (ARMA) [5]. EMAR enables a negotiating agent to acquire an opponent model and to use this model for adjusting its target utility in real time on the basis of an adaptive concession-making mechanism. Hao and Leung proposed a negotiation strategy named ABiNeS, which was introduced for negotiations in complex environments [10]. ABiNeS adjusts the time to stop exploiting the negotiating partner and also employs a reinforcement-learning approach to improve the acceptance probability of its proposals. Williams et al. proposed a novel negotiating agent based on Gaussian processes in multi-issue automated negotiation against unknown opponents [17].

Kawaguchi et al. proposed a strategy for compromising the estimated maximum value based on estimated maximum utility [13]. These papers have been important contributions for bilateral multi-issue closed negotiation; however, they don't deal with multi-times negotiation with learning and reusing the past negotiation sessions. After that, Fujita proposed a compromising strategy with adjusting the speed of making agreements using the Conflict Mode, and focused on multi-times negotiations. However, these strategies focused on the linear utility function only. In real life, most utility functions are nonlinear because of the complexity of the preference structures. Most existing negotiation protocols, though well suited for linear utility functions, work poorly when applied to nonlinear problems because of the complexity of utility domain, multiple optima, and interdependency between issues. However, the negotiation strategy based on the compromising strategy by Fujita [7, 8] can adapt to a nonlinear situation. In this chapter, we demonstrate that the novel negotiation strategy based on the compromising strategy is effective in nonlinear domains, not only the linear domain.

3 Negotiation Environments

The interaction between negotiating parties is regulated by a *negotiation protocol* that defines the rules of how and when proposals can be exchanged. The competition used the alternating-offers protocol for bilateral negotiation as proposed in [16], in which the negotiating parties exchange offers in turns. The alternating-offers protocol conforms with our criterion to have simple rules. It is widely studied in the literature, both in game-theoretic and heuristic settings of negotiation. For example, *Agents A* and *B* take turns in the negotiation. One of the two agents is picked at random to start. When it is the turn of agent *X* (*X* being *A* or *B*), that agent is informed about the action taken by the opponent. In negotiation, the two parties take turns selecting the next negotiation action. The possible actions are:

- *Accept*: It indicates that the agent accepts the opponent's last bid, and the utility of that bid is determined in the utility spaces of agents *A* and *B*.
- *Offer*: It indicates that the agent proposes a new bid.
- *End Negotiation*: It indicates that the agent terminates the entire negotiation, resulting in the lowest possible score for both agents.

If the action was an *Offer*, agent X is subsequently asked to determine its next action, and the turn-taking goes to the next round. If it is not an *Offer*, the negotiation has finished. The turn-taking stops and the final score (utility of the last bid) is determined for each of the agents as follows. The action of agent X is an Accept. This action is possible only if the opponent actually made a bid. The last bid of the opponent is taken, and the utility of that bid is determined in the utility spaces of agents A and B. When the action is returned as EndNegotiation, the score of both agents is set to the lowest score.

The parties negotiate over *issues*, and every issue has an associated range of alternatives or *values*. A negotiation outcome consists of a mapping of every issue to a value, and the set Ω of all possible outcomes is called the negotiation *domain*. The domain is common knowledge to the negotiating parties and stays fixed during a single negotiation session. Both parties have certain preferences prescribed by a *preference profile* over Ω. These preferences can be modeled by means of a utility function U that maps a possible outcome $\omega \in \Omega$ to a real-valued number in the range [0, 1]. In contrast to the domain, the preference profile of the players is private information.

An agent's utility function in the formulation is described in terms of constraints. There are l constraints, $c_k \in C$. Each constraint represents a region in the contract space with one or more dimensions and an associated utility value. In addition, c_k has value $v_a(c_k, \mathbf{s})$ if and only if it is satisfied by contract $\mathbf{s}$. Every agent has its own, typically unique, set of constraints. An agent's utility for contract $\mathbf{s}$ is defined as the weighted sum of the utility for all the constraints it satisfies, i.e. as $u_a(\mathbf{s}) = \sum_{c_k \in C, \mathbf{s} \in x(c_k)} v_a(c_k, \mathbf{s})$, where $x(c_k)$ is a set of possible contracts (solutions) of c_k. This expression produces a "bumpy" nonlinear utility function with high points where many constraints are satisfied and lower regions where few or no constraints are satisfied. This represents a crucial departure from previous efforts on multi-issue negotiation, where contract utility is calculated as the weighted sum of the utilities

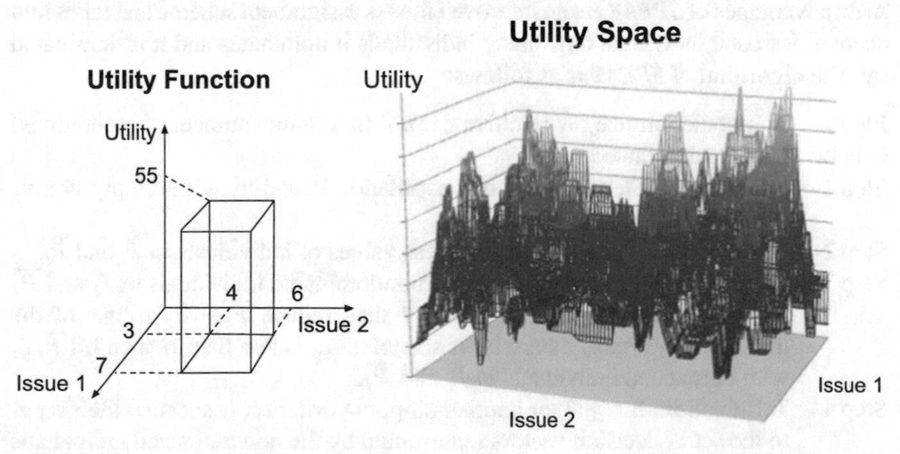

Fig. 1 Example of nonlinear utility space

for individual issues, producing utility functions shaped like flat hyperplanes with a single optimum.

Figure 1 shows an example of a utility space generated via a collection of binary constraints involving Issues 1 and 2. In addition, the number of terms is two. The example, which has a value of 55, holds if the value for Issue 1 is in the range [3, 7] and the value for Issue 2 is in the range [4, 6]. The utility function is highly nonlinear with many hills and valleys. This constraint-based utility function representation allows us to capture the issue interdependencies common in real-world negotiations. The constraint in Fig. 1, for example, captures the fact that a value of 4 is desirable for issue 1 if issue 2 has the value 4, 5 or 6. Note, however, that this representation is also capable of capturing linear utility functions as a special case (they can be captured as a series of unary constraints). A negotiation protocol for complex contracts can, therefore, handle linear contract negotiations.

A negotiation lasts a predefined time in seconds (*deadline*). The timeline is normalized, i.e. time $t \in [0, 1]$, where $t = 0$ represents the start of the negotiation and $t = 1$ represents the deadline. Apart from a deadline, a scenario may also feature discount factors, which decrease the utility of the bids under negotiation as time passes. Let d in [0, 1] be the discount factor. Let t in [0, 1] be the current normalized time, as defined by the timeline. We compute the discounted utility U_D^t of an outcome ω from the undiscounted utility function U as follows: $U_D^t(\omega) = U(\omega) \cdot d^t$. At $t = 1$, the original utility is multiplied by the discount factor. Furthermore, if $d = 1$, the utility is not affected by time, and such a scenario is considered to be undiscounted.

4 SPEA2 for Finding Pareto Frontier

SPEA2 (Strength Pareto Evolutionary Algorithm 2) is the genetic algorithm for finding the pareto frontier in multi-objective optimization proposed by Zitzler [19]. The main advantages of *SPEA2* are an improved fitness assignment scheme that takes into account for each individual how many individuals it dominates and it is dominated by. The algorithm of *SPEA2* is as follows:

Input: N (population size), $\overline{N}$ (archive size), T (maximum number of generations)
Output: A (nondominated set)
Step 1 Initialization: Generate an initial population P_0 and create the empty archive (external set) $\overline{P}_0$. Set $t = 0$.
Step 2 Fitness assignment: Calculate fitness values of individuals in P_t and $\overline{P}_t$.
Step 3 Environmental selection: Copy all nondominated individuals in P_t and $\overline{P}_t$ to $\overline{P}_{t+1}$. If size of $\overline{P}_{t+1}$ exceeds $\overline{N}$ then reduce $\overline{P}_{t+1}$ by means of the truncation operator; otherwise if size of $\overline{P}_{t+1}$ is less than $\overline{N}$ then fill $\overline{P}_{t+1}$ with dominated individuals in P_t and $\overline{P}_t$.
Step 4 Termination: If $t \geq T$ or another stopping criterion is satisfied then set A to the set of decision vectors represented by the nondominated individuals in $\overline{P}_{t+1}$. Stop.

Step 5 Mating selection: Perform binary tournament selection with replacement on $\overline{P}_{t+1}$ in order to fill the mating pool.

The fitness assignment of *SPEA2* is as follows. First, the number of individuals $S(i)$ dominated the individual i is calculated. $R(i)$(raw fitness of the individual i) is the sum of the $S(j)$ with the individual j dominates the individual i. $R(i) = 0$ corresponds to a non-dominated individual, while a high $R(i)$ value means that i is dominated by many individuals. The $\sqrt{N + \overline{N}}$-th element gives the distance sought, denoted as σ_i. The density $D(i)$ is defined by $D(i) = \frac{1}{\sigma_i+2}$. The fitness assignment $F(i)$ is defined by $F(i) = R(i) + D(i)$.

In *SPEA2*, the individual means the proposed bid, and the genetic locus means the issue. The values of each issue mean the genes. The crossover is used as the uniform crossover.

5 Strategy Considering Interdependency Between Issues

5.1 Estimating Opponent's Utility Considering Interdependency Between Issues

The proposed method estimates the opponent's utility based on its statistical information by dividing the opponent's bids into some small elements. In the alternative offering protocol, the bids proposed by the opponent many times should be considered as the important and highly valued ones. However, it is hard to get statistical information by counting all the bids simply because the proposed bids are limited in one-shot negotiation. In addition, we can't estimate effectively in the nonlinear utility functions if the issue interdependencies are ignored. Therefore, we propose a novel method in which the opponent's bid is divided into small elements considering the combinations between issues, and the number of the opponent's proposals is counted to estimate the opponent's utility function.

We assume that the alternative with M issues $\mathbf{s}$ is divided into small parts of the alternative with D elements. $C(D, \mathbf{s})$ is defined as a function of outputting the set of combining D elements from the alternative with M elements ($|C(D, \mathbf{s})| = {}_M C_D$). $count(e)$ returns the number of the element e outputted by $C(D, \mathbf{s})$ in the previous proposals. The estimating function of the opponent's utility of $\mathbf{s}$ is defined as follows:

$$U(\mathbf{s}) = \sum_{e \in C(D,\mathbf{s})} count(e). \tag{1}$$

Figure 2 shows an example of dividing the proposed bids, when $\mathbf{s}' = (2, 4, 1)$ is proposed by the opponent and some small parts of alternatives in *(the number of divided elements)* $= 2$, $C(2, \mathbf{s}') = \{[i_1 = 2, i_2 = 4], [i_1 = 2, i_3 = 1], [i_2 = 4, i_3 = 1]\}$ are divided. We assume that the previous situation was $count([i_1 =$

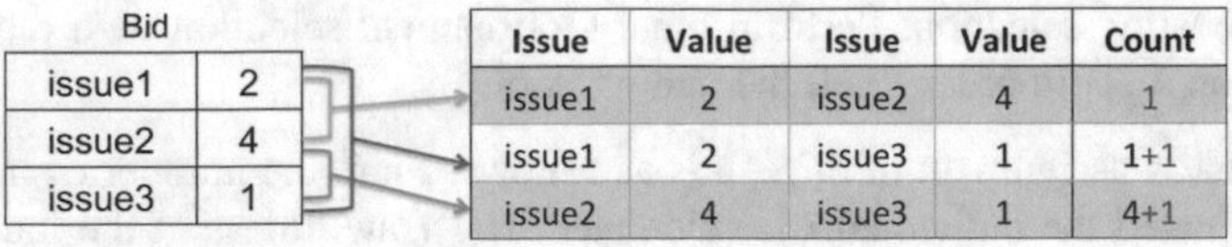

Bid	
issue1	2
issue2	4
issue3	1

Issue	Value	Issue	Value	Count
issue1	2	issue2	4	1
issue1	2	issue3	1	1+1
issue2	4	issue3	1	4+1

Fig. 2 Division of bid and counting method

$2, i_3 = 1]) = 1, count([i_2 = 4, i_3 = 1]) = 4$ before the opponent proposes. In this situation, $count([i_1 = 2, i_2 = 4]) = 1, count([i_1 = 2, i_3 = 1]) = 2, count([i_2 = 4, i_3 = 1]) = 5$ after this proposal is reflected.

After estimating the opponent's utility, our agent can calculate the pareto frontier using our utility function and the opponent's estimated utilities by *SPEA2*.

5.2 *Automated Negotiating Agent Considering Issue Interdependency*

We propose a novel strategy by estimating the opponent's utility function and *SPEA2* proposed in the previous section. The proposed agent has two steps: information-gathering and compromising.

Information-gathering step Early on, the agent initiates the information-gathering step. In this step, the agent doesn't accept the opponent's offers and proposes selfish bids to gather the opponent's information for estimating the opponent's utility. Concretely speaking, the agent proposes the highest 50 bids by searching the genetic algorithm.

Compromising step In the compromising step, the agent proposes the pareto optimal bids estimated by *SPEA2* using the opponent's bids collected by the previous step in order to make agreements. The details are as follows.

First, the agent estimates the pareto frontier using *SPEA2* by considering its own and the opponent's estimated utility function. In moving from the information-gathering step to the compromising step, the agent searches by *SPEA2*, whose generation number is 50. After that, the agent updates a generation from the previous search result using the estimated utility function using the proposed bids by the opponent. The computational time becomes short by considering the differences of the previous proposal.

The agent decides the next actions based on the estimated pareto frontier. The equations for deciding the actions are as follows:

$$u_a(t) = (1 - t^d)(1 - p_a(t)) + p_a(t) \tag{1}$$

$$target_a(t) = \min_{t' \leq t} u_a(t'). \tag{2}$$

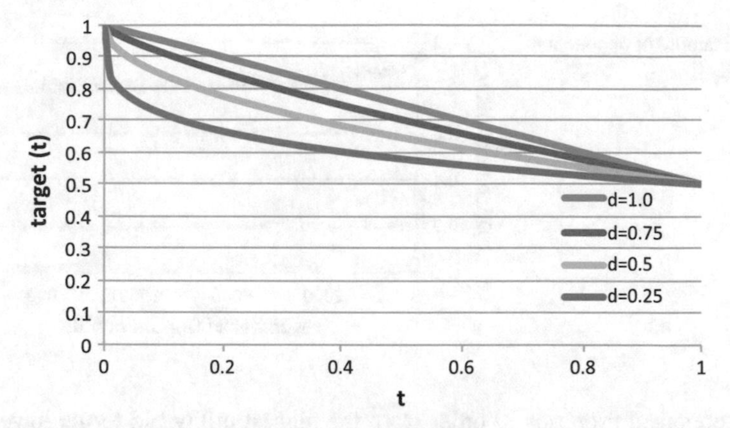

Fig. 3 Example of threshold of offer and accept ($p(t) = 0.5$)

$p_a(t)$ is defined as the lowest utility of the agent in the estimated pareto frontier when the timeline is t. In other words, $p_a(t)$ means the bids with the highest utility in the estimated pareto frontier. When the agent is A and the opponent is B, $p_A(t) = u(\mathbf{s}')\{\mathbf{s}' \mid \max_{\mathbf{s}} u_B(\mathbf{s}) \cap \mathbf{s}\, satisfies\ the\ pareto\ frontier\}$. Next, the agent proposes a bid that satisfies the pareto frontier, and utility is more than $target_a(t)$.

Figure 3 shows the changes of $target_a(t)$ when $p_a(t) = 0.5$. As the discount factor d is small, the agent compromises with the opponent in the earlier stage. By deciding the proposal and acceptance strategies using $p_a(t)$, the agent can compromise considering the rate of conflict.

When the utility of the opponent's proposal in our utility function is higher than $target_a(t)$, the offer is accepted. In addition, our agent terminates this negotiation when it can't gain more than the reservation value; in other words, $target_a(t)$ is less than the reservation value.

6 Experimental Results

6.1 Finding the Pareto Optimal Bids

We conducted several experiments to evaluate our approach. The following parameters were used. The domain for the issue values was [0, 9]. The number of issues is from 10 to 30 at 5-issue intervals, and all experiments are conducted in five different domains. The number of constraints was (*the number of issues*) × *10*, and these constraints are related to 3.5 issues on average. The value for a constraint was decided randomly from 0 to 1.0. The following constraints would all be valid: Issue 1 = [2, 6], Issue 3 = [2, 9]. The utility functions of each agent are decided randomly.

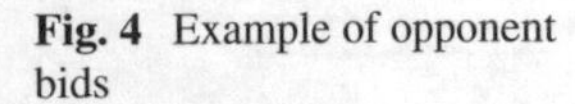

Fig. 4 Example of opponent bids

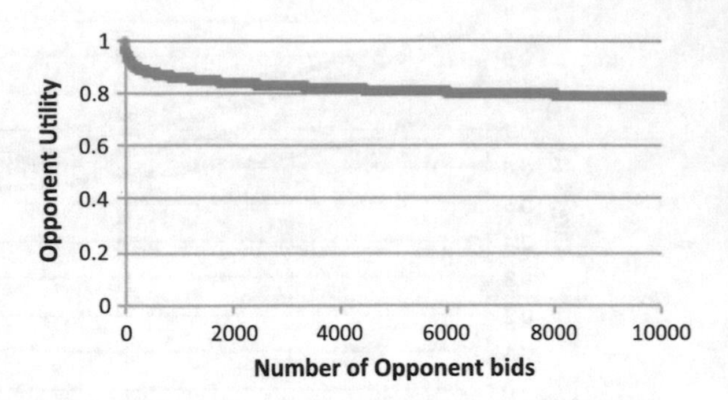

The opponent proposes in order from the highest utility bid to the lower ones. Figure 4 shows the opponent's proposals in the experiments when the number of the opponent's bids increases. As Fig. 4 shows, the opponent proposes a bid as a gradual compromise by selecting bids with lower utilities. In the initial phase, most of the negotiating agents wait and see their opponent's bids before judging their opponent's strategy and utilities in the previous ANAC [1]. Negotiating agents usually propose selfish bids to avoid such mistakes as lower utility for their side in the initial phase. After that they propose a bid as a gradual compromise by selecting bids with lower utilities for themselves. Therefore, the opponent's strategy in these experiments is created as in Fig. 4 to evaluate in the previous virtual ANAC.

In the experiments, we evaluated the quality of the pareto frontier in our approach measured by the size of the dominant area [18]. The dominant area increases when the solution set is close to the pareto frontier and exists without any spaces between solutions. This evaluation measure uses the size of the dominant area in the experiments because it is effective for our approach to finding the pareto frontier. The numbers of combinations D are 1, 2 or 3 in the experiments. In *SPEA2*, the number of individuals is 250, the archive size is 250, the mutational rate is *1/(the number of issues)*, and 500,000 function evaluations.

The optimal pareto frontier was found by searching its own and the 'opened' opponent's utility spaces with *SPEA2*. Optimality Rate is defined as *(dominant area achieved by each protocol)/(optimal dominant area calculated by SPEA2)*.

Figure 5 shows the average of the optimality rate in all utility spaces when the number of the opponent's bids changes. As the number of the opponent's bids increases, our approach can find better solutions that are close to the pareto frontier. However, the average of the dominant area is almost the same when the number of opponent's bids exceeds 10,000. The reason for this is that the opponent proposes the low-valued bids in the final phase; therefore, these bids don't have relations with finding the pareto frontier.

Figure 6 shows the average of the optimality rate in all utility spaces when the number of combinations (D) changes. Our approach of $D = 2$ and $D = 3$ can search the better bids that are closer to the pareto frontier than that of $D = 1$. Estimated values of our approach of $D = 1$ are calculated by dividing bids with each issue without considering the interdependency between issues. On the other hand, the

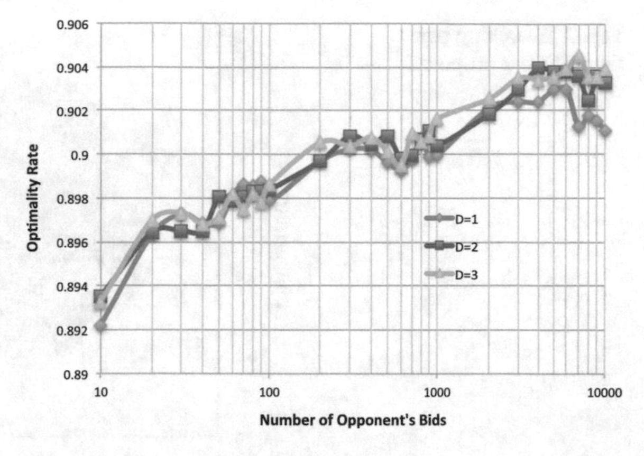

Fig. 5 Dominant area with number of opponent's bids

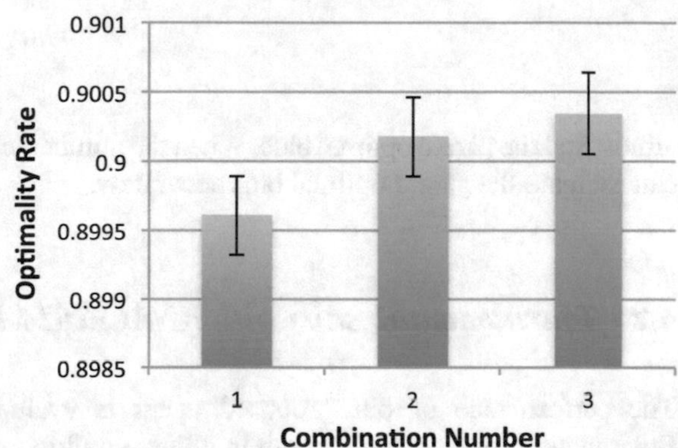

Fig. 6 Dominant area with combination number

approach needs to have a power of expressions to the nonlinear utility functions because some issues have interdependency between issues in the nonlinear utility functions. Therefore, our approach of $D = 2$ and $D = 3$ can have better results than that of $D = 1$. Moreover, our approach of $D = 3$ was better than that of $D = 2$ in this experiment. In other words, our approach can improve the performance as the number of issues considering the interdependency increases. On the other hand, the computation time of our approach increases exponentially as the number of combinations (D) increases. When the number of issues M is enough larger than the number of combinations D, the time complexity is $O(M^D)$. In other words, the number of combinations (D) is important for our approach to controlling the computational complexity.

Figure 7 shows the results of our proposed approach when the number of issues is 6. The horizontal axis is the utility of our agent, and the vertical axis is the utility of the opponent. "Perfect" means the pareto frontier when the utility spaces of our side and the opponent's side are opened. "Estimate" means the pareto frontier estimated by our proposed approach after the negotiation. As Fig. 7 shows, our approach can

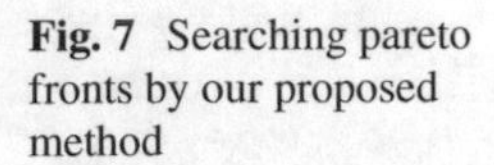

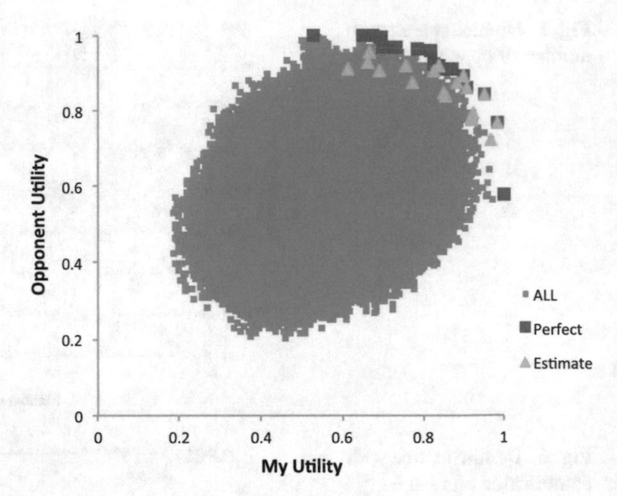

Fig. 7 Searching pareto fronts by our proposed method

almost find the pareto optimal bids. When the number of issues is larger, our approach can estimate the pareto optimal bids accurately.

6.2 *Tournament Results with ANAC-2014 Finalists*

The performance of our proposed agent is evaluated with GENIUS (General Environment for Negotiation with Intelligent multipurpose Usage Simulation [15]), which is also used as a competition platform for ANAC. Ten agents were selected from the qualifying round of ANAC-2014 competition: AgentM, Gangster, WhaleAgent, E2Agent, AgentYK, BraveCat v0.3, GROUP2Agent, kGAgent, DoNA, ANAC2014Agent.[1]

The domains were selected from archives of ANAC-2014. The sizes of the domains are $10^{10}, 10^{30}, and\ 10^{50}$. Each constraint in these domains is related to 1 to 5 issues. The horizontal axis means agent A's utility and the vertical axis means agent B's utility in each figure. The scenarios contained broadly similar characteristics such as the shapes of the pareto frontier and so on. In all domains, the discount factors are set to 1.0 and 0.5, and the reservation values are set to 0 and 0.75, respectively. For each pair of agents, under each utility function, we ran a total of 12 negotiations (including the exchange of preference profiles). The maximum negotiation time of each negotiation session is set to 3 min and normalized into the range of [0, 1]. Figures 8 and 9 show the mean scores over all the individual welfares and social welfares achieved by the round-robin tournament among 11 agents (our agent and 10 state-of-the-art agents in ANAC-2014).

[1] All agents and domains participating in the final round of ANAC-2014 are available in the newest GENIUS.

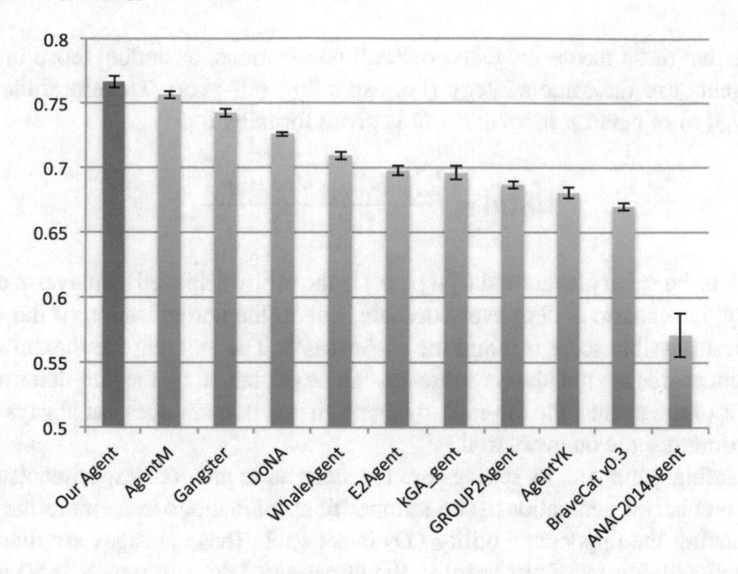

Fig. 8 Average of individual utilities among ANAC-2014 finalists (The *lines* show standard errors)

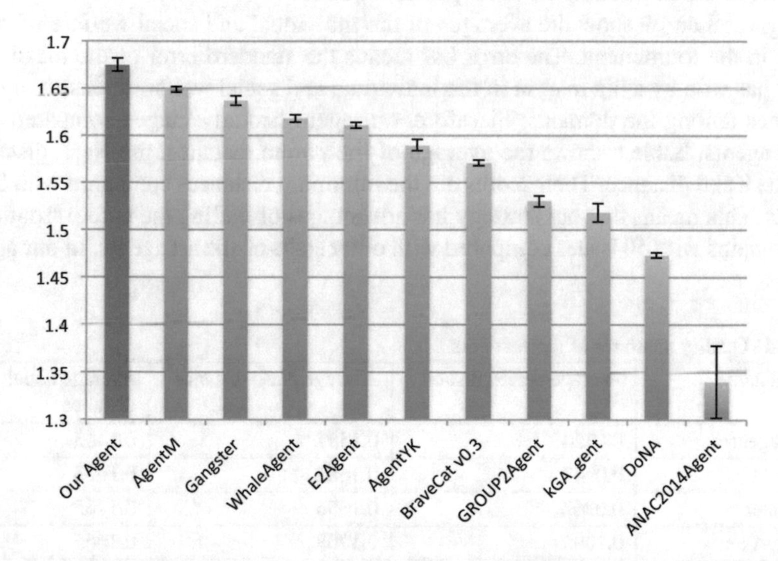

Fig. 9 Average of social welfares among ANAC-2014 finalists (The *lines* show standard errors)

Note that these means are taken over all negotiations, excluding those in which both agents use the same strategy (i.e. excluding self-play). Therefore, the mean score $U_\Omega(p)$ of agent p in scenario Ω is given formally by:

$$U_\Omega(p) = \frac{\sum_{p' \in P, p \neq p'} U_\Omega(p, p')}{(|P| - 1)}, \tag{3}$$

where P is the set of players and $U_\Omega(p, p')$ is the utility achieved by player p against player p' in scenario Ω. For every domain, due to the normalization of the scores, the lowest possible score is 0 and the highest is 1. The fact that the maximum and minimum scores are not always achieved can be explained by the non-deterministic behavior of the agents: the top-ranking agent in one domain does not always obtain the maximum score on every trial.

Regarding our agent, its settings are the same as in previous experiments except for the number of generations. The number of combinations in dividing the issues in estimating the opponent's utility (D) is set to 2. These changes are due to the computational powers of our agent in the larger-sized domains (such as 50 issues). In the larger-sized domains, our agent can't make enough proposals by the time limit because of the large computational power required.

Figures 8 and 9 show the averages of the individual and social welfare of every agent in the tournament. The error bar means the standard error of the mean. Our agent has won by a big margin in the individual and social welfares considering the variance among the domains; therefore, our agent had advantages compared with other agents. Table 1 shows the average of the pareto distance, the Nash distance, and the Kalai distance. Their scores are the minimum distances compared with other agents. This means that our strategy has advantages of finding the pareto frontier in the domains with 50 issues compared with other state-of-the art agents. In our agent,

Table 1 Quality measures of tournaments

Agent name	Average pareto distance	Average Nash distance	Average Kalai distance
Our Agent	0.0724	0.1557	0.1465
AgentM	0.0942	0.1682	0.1623
Gangster	0.0962	0.1856	0.1765
WhaleAgent	0.1095	0.1768	0.1665
E2Agent	0.1149	0.1987	0.1874
AgentYK	0.1281	0.2094	0.1963
BraveCat v0.3	0.1415	0.2119	0.1994
GROUP2Agent	0.1701	0.2580	0.2520
kGAgent	0.1791	0.2574	0.2498
DoNA	0.2072	0.2831	0.2720
ANAC2014Agent	0.2990	0.3670	0.3596

the combinations of the estimating method of the opponent's preferences by dividing the issues and *SPEA2* have a good effect in finding the pareto frontier effectively, despite that our acceptance and proposal strategies are simple compared with other agents.

7 Conclusions

In this chapter, we proposed an estimating method for the pareto frontier based on an opponent's bids. In the proposed method, the opponent's bid was divided into small elements considering the combinations between issues, and the number of the opponent's proposals was counted to estimate the opponent's utility function. In addition, *SPEA2*, which was based on the genetic algorithm for multiple objective optimization, was employed in our proposed method to search the pareto optimal bids. In the experiments, we evaluated the quality of the pareto frontier in our approach measured by the size of the dominant area. The experimental results demonstrated that our proposed method considering the interdependency between issues can search the pareto optimal bids. In addition, we compared the negotiation efficiency of our proposed agent with ten state-of-the-art negotiation agents that entered the final round of ANAC-2014. Our agent won by a big margin in the negotiation tournaments because of the estimated opponent's utility and *SPEA2*.

Future works will address improvements in estimating the opponent's utility in our proposed approach. To solve this problem, our approach needs to consider the order of an opponent's proposals in estimating the opponent's utility. Another important task is to judge the opponent's strategy based on modeling or machine learning technique to further enhance our proposed method.

References

1. R. Aydogan, T. Baarslag, K. Fujita, T. Ito, C. Jonker, The fifth international automated negotiating agents competition (anac2014) (2014), http://www.itolab.nitech.ac.jp/ANAC2014/
2. T. Baarslag, K. Fujita, E. Gerding, K. Hindriks, T. Ito, N.R. Jennings, C. Jonker, S. Kraus, R. Lin, V. Robu, C. Williams, Evaluating practical negotiating agents: results and analysis of the 2011 international competition. Artif. Intell. J. (AIJ) **198**, 73–103 (2013)
3. T. Baarslag, K.V. Hindriks, Accepting optimally in automated negotiation with incomplete information, in *Proceedings of the 2013 International Conference on Autonomous Agents and Multi-agent Systems* (International Foundation for Autonomous Agents and Multiagent Systems, 2013), pp. 715–722
4. S. Chen, G. Weiss, An efficient and adaptive approach to negotiation in complex environments, in *Proceedings of the 19th European Conference on Artificial Intelligence (ECAI-2012)*, vol. 242 (2012), pp. 228–233
5. S. Chen, G. Weiss, An efficient automated negotiation strategy for complex environments, in *Engineering Applications of Artificial Intelligence* (2013)

6. S.S. Fatima, M. Wooldridge, N.R. Jennings, Multi-issue negotiation under time constraints, in *Proceedings of the First International Joint Conference on Autonomous Agents and Multiagent Systems (AAMAS 2002)* (New York, NY, USA, 2002), pp. 143–150
7. K. Fujita, Automated strategy adaptation for multi-times bilateral closed negotiations, in *Proceedings of the 13th International Conference on Autonomous Agents and Multiagent Systems (AAMAS 2014)* (2014), pp. 1509–1510
8. K. Fujita, Efficient strategy adaptation for complex multi-times bilateral negotiations, in *7th IEEE International Conference on Service-Oriented Computing and Applications (SOCA 2014)* (2014), pp. 207–214
9. K. Gal, T. Ito, C. Jonker, S. Kraus, K. Hindriks, R. Lin, T. Baarslag, The forth international automated negotiating agents competition (anac2013) (2013), http://www.itolab.nitech.ac.jp/ANAC2013/
10. J. Hao, H.-F. Leung, Abines: an adaptive bilateral negotiating strategy over multiple items, in *2012 IEEE/WIC/ACM International Conferences on Intelligent Agent Technology (IAT-2012),*, vol. 2 (2012), pp. 95–102
11. S. Kakimoto, K. Fujita, Estimating pareto fronts using issue dependency for bilateral multi-issue closed nonlinear negotiations, in *7th IEEE International Conference on Service-Oriented Computing and Applications (SOCA 2014)* (2014), pp. 289–293
12. S. Kakimoto, K. Fujita, Randomdance: compromising strategy considering interdependencies of issues with randomness, *Modern Approaches to Agent-based Complex Automated Negotiation (This book)*, Studies in Computational Intelligence (Springer, Berlin, 2016)
13. S. Kawaguchi, K. Fujita, T. Ito, Compromising strategy based on estimated maximum utility for automated negotiation agents competition (anac-10), in *24th International Conference on Industrial Engineering and Other Applications of Applied Intelligent Systems (IEA/AIE-2011)* (2011), pp. 501–510
14. M. Klein, P. Faratin, H. Sayama, Y. Bar-Yam, Negotiating complex contracts. Group Decis. Negot. **12**(2), 58–73 (2003)
15. R. Lin, S. Kraus, T. Baarslag, D. Tykhonov, K. Hindriks, C.M. Jonker, Genius: an integrated environment for supporting the design of generic automated negotiators, in *Computational Intelligence* (2012)
16. A. Rubinstein, Perfect equilibrium in a bargaining model. Econometrica **50**(1), 97–109 (1982)
17. C.R. Williams, V. Robu, E.H. Gerding, N.R. Jennings, Using gaussian processes to optimise concession in complex negotiations against unknown opponents, in *Proceedings of the 22nd International Joint Conference on Artificial Intelligence (IJCAI-2011)* (2011), pp. 432–438
18. E. Zitzler, L. Thiele, Multiobjective evolutionary algorithms: a comparative case study and the strength pareto approach. IEEE Trans. Evol. Comput. **3**(4), 257–271 (1999)
19. E. Zitzler, M. Laumanns, L. Thiele, E. Zitzler, E. Zitzler, L. Thiele, L. Thiele, Spea2: improving the strength pareto evolutionary algorithm (2001)

A Negotiation-Based Model for Policy Generation

Jieyu Zhan, Minjie Zhang, Fenghui Ren and Xudong Luo

Abstract In traditional policy generation models, the preferences over polices are often represented by qualitative orderings due to the difficulty of acquisition of accurate utility. Thus, it is difficult to evaluate agreements in such models so that players cannot adjust their strategies during a policy generation process. To this end, this paper introduces a negotiation-based model for policy generation, which contains two evaluation methods, both from the perspectives of concessional utilities and consistency, to guide players to make decisions flexibly. The first method is used to model humans' reasoning about how to calculate concessional utilities from uncertain preference information of policies based on fuzzy reasoning, while the second method is used to measure similarity between an ideal agreement and an offer based on a prioritised consistency degree. The experimental results show the difference between the evaluation methods and confirm that the proposed model and evaluation methods can help players achieve better agreements than an existing model.

Keywords Automatic multi-issue negotiation · Preference representation · Evaluation · Fuzzy reasoning · Consistency

J. Zhan · X. Luo
Institute of Logic and Cognition, Sun Yat-sen University, Guangdong, China
e-mail: zhanjieyu@gmail.com

X. Luo
e-mail: luoxd3@mail.sysu.edu.cn

J. Zhan · M. Zhang · F. Ren (✉)
School of Computing and Information Technology, University of Wollongong, Wollongong, NSW, Australia
e-mail: fren@uow.edu.au

M. Zhang
e-mail: minjie@uow.edu.au

© Springer International Publishing AG 2017

K. Fujita et al. (eds.), *Modern Approaches to Agent-based Complex Automated Negotiation*, Studies in Computational Intelligence 674,
DOI 10.1007/978-3-319-51563-2_7

1 Introduction

A policy is a statement of intent or a principle of action to guide decisions and achieve certain goals, and policy generation is about how to aggregate voters' opinions or preferences to reach acceptable policies. Policy generation is an important behaviour in our political life and business, which has been studied in different areas, such as political science, management science and economics [1]. People often use qualitative orderings rather than utility functions to represent preferences of different policies [12]. For example, when voters are voting some candidates, it is easy for them to make a preference ordering among candidates rather than measuring the preference difference in a numeric scale [2]. However, there are three disadvantages in most of the studies of policy generation based on social choice theory [1]. (i) In real life, policy generation is so complex that a single preference ordering cannot represent players' personalities well. (ii) There is a lack of interaction in traditional methods such as voting [4]. Policy generation is an interactive process, so players should be able to change their opinions with the reveal of information. (iii) There are few effective methods for evaluating agreements due to qualitative preference orderings [11]. To address the above problems, researchers have been discovering other models to represent policy generation in real life situations. Negotiation models are one of the most suitable models since a negotiation is a process in which a group of parties exchange information to obtain a mutually acceptable outcome [3].

The first two problems have been solved by some existing negotiation models [11, 12]. However, so far the third problem of evaluation of agreements has not been solved well yet. It is significant to give suitable evaluation methods because players cannot make decision well without an evaluation. The above models are all ordinal models, where the utilities of negotiators are not measured by values, but represented by preference orderings over policies. This is the main reason resulting in the evaluation problem. The ordinal model depicts the preference relation between policies intuitively, but it is difficult to evaluate an agreement due to the lack of a qualitative assessment. Therefore, an agreement in an ordinal model is often accord with some certain axioms [5, 11, 12], but may not always be an optimal outcome. Humans can estimate which agreement is better than others but it is hard to say why, because of the complex reasoning rather than uncertain utilities of policies. Thus, an approach is required to model humans' reasoning of how to evaluate an agreement from uncertain preference information about policies. To this end, we construct a new negotiation model for policy generation, in which we propose two kinds of methods for the outcome evaluations to guide players to make good decisions.

The proposed research advances the state of art in the field of negotiation-based policy generation models in the following aspects. (i) An agent model is proposed, which describes a negotiation scenario intuitively. (ii) Based on the agent model, a fuzzy reasoning based method is proposed to calculate agents' concessional utilities, which is used to help agents to evaluate their outcomes or offers in the case where the utilities of policies are not represented by precise numbers. (iii) A prioritised consistency method is introduced to evaluate the similarity between a player's ideal state and an agreement (or an offer).

The rest of this paper is organised as follows. Section 2 introduces our negotiation-based model. Section 3 presents the fuzzy reasoning, which is used to calculate the concessional utility in the utility-based evaluation method. Section 4 proposes a priority operator to calculate the prioritised consistency of an agreement in the consistency-based evaluation method. Section 5 illustrates our model by an example. Section 6 demonstrates the experimental results and the analysis. Section 7 discusses the related work. Finally, Sect. 8 concludes the paper with future work.

2 A Negotiation-Based Model

This section presents a negotiation-based policy generation model. Our model focuses on the negotiation environment where negotiators have different attitudes to some policies but need to find out an acceptable agreement, and they have no accurate utilities of policies. Firstly, we specify a policy structure. Secondly, we propose two evaluation methods of offers. Thirdly, we introduce the negotiation protocol for generating policies. Finally, we define two kinds of agreement concepts.

2.1 A Policy Structure

In our model, agents can propose policies they are interested in and express their opinions on these policies (support or opposition). Policies can be represented by propositions. For example, *disarmament* is a policy and *joining a military alliance* is also a policy. We describe a policy through two dimensions: one is about agent's attitude to the policy, while the other is about how important the policy is to the agent, which denoted as a preference degree in this paper. Formally, we have:

Definition 1 A policy structure $\mathscr{S}$ is a tuple of (X, Att, Pre), where:

- X is a finite set of propositions in a propositional language $\mathscr{L}$ and $\forall x \in X$, x is a policy represented by a propositional variable;
- Att is an attitude function defined as $Att : X \to \{0, 1\}$ and $\forall x \in X$, $Att(x)$ is called the agent's attitude of policy x. $Att(x) = 1$ means that the agent supports the policy and $Att(x) = 0$ means that the agent opposes the policy; and
- Pre is a preference function defined as $Pre : X \to (0, 1]$ and $\forall x \in X$, $Pre(x)$ is called the preference degree of x in set X, which represents how important every policy is to the agent.

For a certain policy, an agent should only have an attitude in the beginning, supporting or opposing the policy, and does not have an attitude to any irrelevant policy since those policies do not affect his utility in an agreement. For example, agent i supports the policy *disarmament*, then his attitude to this issue is $Att_i(disarmament) = 1$. In fact, opposing a policy can also be represented by a sentence [5, 11, 12], such as *oppose disarmament*, but we differentiate attitudes and policies in this paper because we aim to deal both policies in continuous domains and discrete domains. In discrete domains, the policy should be totally accepted or rejected by an agent. Take the policy *joining a military alliance* as an example. In the final agreement, such policy is either supported or opposed. However, in continuous domains, a policy can be partially accepted in a final agreement, that is, the acceptance domain is divisible. If the policy is *disarmament*, supporting such a policy may mean to reduce ten thousand soldiers and opposing it may mean to totally maintain the existing size of the army, then in agreement, *disarmament* can be partly accepted and opposed. A successful negotiation can lead to an acceptable level, such as reducing five thousand soldiers. The preference degree is used to represent the importance of a policy to an agent. Instead of giving a numeric utility to value an issue, we use linguistic terms, for instance, *important*, *not important*, to depict the preference of a policy. For a statement "policy A is important", there is a degree of truth, and the fuzzy set theory is used to quantify this degree. In our model, humans need to offer a preference degree between 0 and 1. With membership functions of linguistic terms, we can depict the importance of policies. Although this kind of information about preference is more specific than preference ordering through pairwise comparisons, it is easier to be obtained than numeric utilities.

2.2 Evaluation Methods of an Offer

Our model uses an extension of alternating-offers protocol, so in every round one of the agents gives an offer to the others. Formally, we define an offer as follows:

Definition 2 $O_{i,\lambda}$ is an offer function of agent i defined as $O_{i,\lambda} : X_i \bigcup X_{-i} \to [0, 1]$, where X_{-i} denotes the policy sets of all i's opponents, and $\forall x \in X_i \bigcup X_{-i}$, $O_{i,\lambda}(x)$ is called agent i's acceptance degree of policy x in the λ-th round.

This part proposes two evaluation methods of an offer to guide agents to make decisions. These two evaluation methods are based on different perspectives, one is utility-based method and the other is consistency-based method.

2.2.1 A Utility-Based Method

Giving an offer, fuzzy reasoning is used to evaluate concessional utility for agents. Suppose that if all of the acceptance degrees of an agent's policies in an offer are the same with its attitudes to its policies, then the utility of the agent is the highest and there is no concessional utility. Intuitively, the more concessions an agent makes on a policy, the more utility declines. If the policy is more important to the agent, then the utility will decreases more. We model the concessional utilities as follows:

Definition 3 A concessional utility function of agent i, denoted as $\triangle u_i$, is given by:

$$\triangle u_i = \frac{\sum_{x \in X_i} FR(Pre_i(x), Con_{i,\lambda}(x))}{|X_i|}, x \in \{x | Con_{i,\lambda}(x) \neq 0\}, \quad (1)$$

where $|X_i|$ is the number of policies in policy set X_i and $Con_{i,\lambda}$ is a concession degree function to represent a degree to which a negotiator makes a concession on a policy, which is defined as

$$Con_{i,\lambda}(x) = \left| Att_i(x) - O_{i,\lambda}(x) \right|, \quad (2)$$

where $x \in X_i$ and FR is a kind of fuzzy reasoning based on intuitive fuzzy rules for calculating concessional utility of every policy.

We take the time cost into consideration in this model. The same offer in different rounds has different concessional utilities for an agent. The utility an agent obtains in an offer is lower than the same one in a previous round. The concessional utility becomes larger and larger as time goes on. So, the following concept is introduced:

Definition 4 A concessional utility function with time constrains of agent i, denoted as $\triangle u_i^t$, is given by:

$$\triangle u_i^t(\triangle u_i, \sigma, \lambda) = \triangle u_i^{\sigma^{\lambda-1}}, \quad (3)$$

where $\sigma \in [0, 1]$ is a discount factor, which is used to decrease the utility of the offers as time passes and λ refers to the λth round of the negotiation.

When giving an offer, agents determine whether to accept it or not, so an agent should have an acceptable threshold of concessional utility in every round. That is, if one of the opponents' offer makes a concession utility larger than the threshold, the agent should reject it and vice versa. If the agent rejects an offer, it should generate an offer not larger than the threshold. Formally, we have:

Definition 5 The acceptable threshold of concessional utility of agent i, denoted as $\triangle \hat{u}_i$, is given by:

$$\triangle \hat{u}_i = f(\lambda) \leqslant \triangle u_{i,\max}, \quad (4)$$

where

$$\triangle u_{i,\max} = \frac{\sum_{x \in X_i} FR\,(Pre_i(x), 1)}{|X_i|},$$

$\triangle u_{i,\max}$ means the highest concessional utility of agent i in a negotiation. It happens only when it makes the largest concessions of all policies in its policy set.

2.2.2 A Consistency-Based Method

In our model, a policy in an offer is presented by a proposition with partial truth. So, we can evaluate an offer in the perspective of logical consistency, that is, how similar two policy sets are. Because policies are of varying significance to an agent, we take the preference into consideration when evaluating the similarity of consistency between the original policy set and an offer. The idea is when a policy is less important, the inconsistency of its truth value between the original policy set and the offer does less harm to the consistency between two sets. Formally, we have:

Definition 6 For an offer function O, a prioritised consistency degree ρ of the offer for agent i is given by:

$$\rho_i = \sqrt{\frac{\sum_{x\in X_i}(Pre_i(x)\diamond\gamma_i(x))^2}{\sum_{x\in X_i}(Pre_i(x)\diamond 1)^2}}, x\in X_i, \tag{5}$$

where $\gamma_i(x) = 1 - |O(x) - Att_i(x)|$ is the consistency degree of x for agent i, and operator $\diamond$: $[0, 1] \times [0, 1] \to [0, 1]$ is a priority operator that satisfies the following properties:

(i) $\forall\, a_1, a_2, a_2' \in [0, 1], a_2 \leqslant a_2' \Rightarrow a_1 \diamond a_2 \leqslant a_1 \diamond a_2'$,
(ii) $\forall\, a_1, a_1', a_2 \in [0, 1], a_1 \leqslant a_1' \Rightarrow a_1 \diamond a_2 \geqslant a_1' \diamond a_2$,
(iii) $\forall a \in [0, 1], 1 \diamond a = a$, and
(iv) $\forall a \in [0, 1], 0 \diamond a = 1$

Similarly to the concept of acceptable threshold of concessional utility, we define the acceptable threshold of prioritised consistency as follows:

Definition 7 The acceptable threshold of prioritised consistency degree of agent i, denoted as $\hat{\rho}_i$, is given by:

$$\hat{\rho}_i = g(\lambda) \geqslant \rho_{i,\min}, \tag{6}$$

where

$$\rho_{i,\min} = \sqrt{\frac{\sum_{x\in X_i}(Pre_i(x)\diamond(1-|1|))^2}{\sum_{x\in X_i}(Pre_i(x)\diamond 1)^2}}, \tag{7}$$

$\triangle u_{i,\max}$ denotes the smallest similarity of consistency between the original policy set and an offer of agent i in a negotiation. It happens only when all relevant policies in an offer diametrically opposed to an agent's original attitudes to policies.

A negotiator agent can be formally defined as follows:

Definition 8 An agent i is a tuple of $(\mathscr{S}, (\triangle u^t, \triangle\hat{u}), (\rho, \hat{\rho}))_i$, where $\triangle u^t$, $\triangle\hat{u}$, ρ and $\hat{\rho}$ are defined in the above definitions.

Policy set $\mathcal{S}$ describes agent i's polities and respective attitudes and preference levels. Two-tuples $(\Delta u^t, \Delta\hat{u})$ and $(\rho, \hat{\rho})$ give two ways to evaluate an offer.

2.3 Negotiation Protocol

The agents communicate with each other based on an extension of alternating-offers protocol [9]. That is, the negotiators exchange offers in turn until the negotiation is finished. We extend the bilateral negotiation model to multilateral model for policy generation in real life. More specifically, one of the agents starts the negotiation randomly and the others reply to it. The first offer of every agent should indicate its attitude of the policies it concerns. Formally, the first offer of agent i, denoted as $O_{i,1}$, should satisfy the following property: $\forall x_i \in X_i$, $O_{i,1}(x_i) = Att_i(x_i)$. This means that the agent should reveal his attitude at the beginning, because this kind of offer can maximise its utility. An agent can choose three actions as a reply, including accepting the offer; rejecting the offer with generating a new one to the opponents; and ending the negotiation that resulting in the lowest utilities for all agents. If every negotiator has proposed an offer in turn, then a complete negotiation round is finished and if there is no offer supported by all negotiators, a new round will begin. Agents just know each other's policies and attitudes, but for preventing the exploration of opponents, they do not reveal their preference of policies. Some policies proposed by one agent i may be irrelevant to agent j and different outcomes of this kind of policies do not influence j's utility. We call such policies as irrelevant policies for party j, but an agreement of negotiation should also include such kind of policies.

2.4 Agreement Generation

An agreement appears if and only if an offer proposed by one of the agents is supported by all the other agents. According to different evaluation methods, we distinguish two agreement concepts. The first one is a utility-based agreement. That is, the agents obtain an agreement in the utility scale and evaluate the agreement by concessional utility. The latter one is a consistency-based agreement. That is, the agents find an agreement consistent enough for all of them. Formally, we have:

Definition 9 An offer function O proposed in λth round in a N-agent negotiation is a utility-based agreement if it satisfies: $\forall i \in N$, $\Delta u_i^t(\Delta u_i, \sigma, \lambda) \leqslant \Delta\hat{u}_i(\lambda)$,

Definition 10 An offer function O proposed in λth round in a N-agent negotiation is a consistency-based agreement if it satisfies: $\forall i \in N$, $\rho_i(\lambda) \geqslant \hat{\rho}_i(\lambda)$.

3 Fuzzy Reasoning

In this section, a fuzzy reasoning was proposed to evaluate concessional utilities.

3.1 *Fuzzy Linguistic Terms of Fuzzy Variables*

We define fuzzy set as follows:

Definition 11 Let U be a set (domain). A fuzzy set A on U is characterised by its membership function

$$\mu_A : U \rightarrow [0, 1]$$

and $\forall u \in U$, $\mu_A(u)$ is called the membership degree of u in fuzzy set A.

The concessional utility of policies mainly depends on two factors: the preference degree and the concession degree. So we use fuzzy sets based on the domains of preference and concession. We distinguish various levels of both domains by different linguistic terms. We use four terms, *very important*, *important*, *fairly important* and *less important* to depict different levels of preference of a policy. Similarly, we use five terms, *very high*, *high*, *medium*, *low* and *very low* to represent different levels of concession an agent makes on a policy and indicate different levels of the output of fuzzy reasoning, that is the concessional utility of a policy.

In this paper, we employ the trapezoidal-type fuzzy membership function [8]:

$$\mu(x) = \begin{cases} 0 & \text{if } x \leq a, \\ \frac{x-a}{b-a} & \text{if } a \leq x \leq b, \\ 1 & \text{if } b \leq x \leq c, \\ \frac{d-x}{d-c} & \text{if } c \leq x \leq d, \\ 0 & \text{if } x \geq d. \end{cases} \tag{8}$$

We draw the membership functions of linguistic terms of preference degree, concession degree and concessional utility in Figs. 1, 2 and 3.

3.2 *Fuzzy Rules*

A fuzzy reasoning is based on fuzzy rules (i.e., IF-THEN rules). We show the fuzzy rules in Table 1. Rule 1 denotes that if the preference degree of a policy is less important and the concession degree of the policy is very low, the concessional utility of the policy is very low. Such rule models the intuitive reasoning that if a person makes a little concession on a less important policy, then compared to his original offers, his concessional utility is very low. Similarly, we can understand other rules.

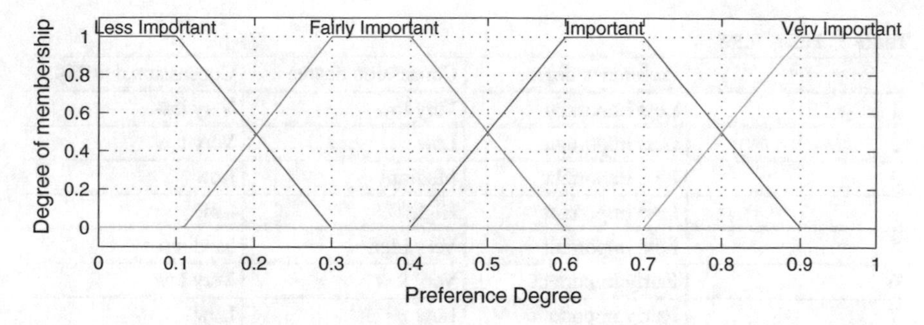

Fig. 1 Membership function of preference degree

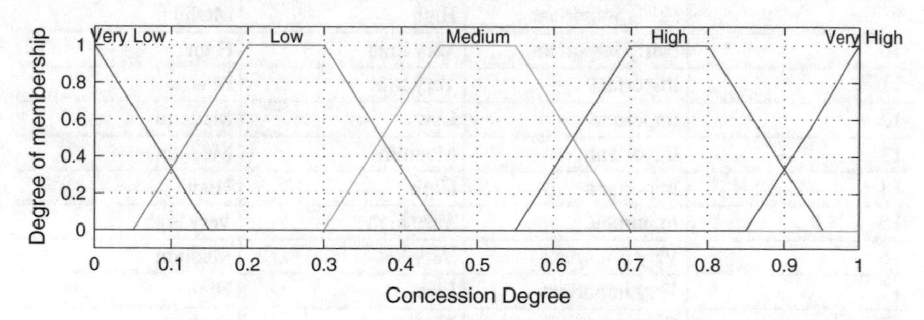

Fig. 2 Membership function of concession degree

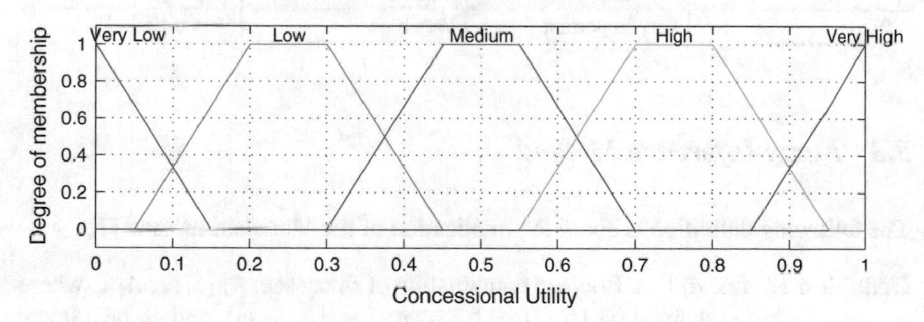

Fig. 3 Membership function of concessional utility

These rules and the linguistic terms of inputs and output may vary among different persons. For example, some persons use more terms to depict the different levels of preference degrees, while the others may consider that two terms, *important* and *less important*, are enough. In our model, the linguistic terms, membership functions and fuzzy rules depend on particular persons, thus humans can adjust them in fuzzy reasoning in their agents to model their reasoning. In this paper, we adopt the above setting to interpret our model.

Table 1 Fuzzy rules

	Preference degree	Concession degree	Concessional utility
1	Less important	Very low	Very low
2	Less important	Low	Very low
3	Less important	Medium	Low
4	Less important	High	Low
5	Less important	Very high	Medium
6	Fairly important	Very low	Very low
7	Fairly important	Low	Low
8	Fairly important	Medium	Low
9	Fairly important	High	Medium
10	Fairly important	Very high	High
11	Important	Very low	Low
12	Important	Low	Medium
13	Important	Medium	Medium
14	Important	High	High
15	Important	Very high	Very high
16	Very important	Very low	Medium
17	Very important	Low	High
18	Very important	Medium	High
19	Very important	High	Very high
20	Very important	Very high	Very high

3.3 Fuzzy Inference Method

The following definition is about the implication of the Mamdani method [7].

Definition 12 Let A_i be a Boolean combination of fuzzy sets $A_{i,1}, \ldots, A_{i,m}$, where $A_{i,j}$ is a fuzzy set defined on $U_{i,j}$ $(i = 1, \ldots, n;\ j = 1, \ldots, m)$, and B_i be a fuzzy set on U' $(i = 1, \ldots, n)$. Then when the inputs are $\mu_{A_{i,1}}(u_{i,1}), \ldots, \mu_{A_{i,m}}(u_{i,m})$, the output of such fuzzy rule $A_i \rightarrow B_i$ is fuzzy set B'_i defined as follows: $\forall u' \in U'$,

$$\mu_i(u') = \min\{f(\mu_{A_{i,1}}(u_{i,1}), \ldots, \mu_{A_{i,m}}(u_{i,m})), \mu_{B_i}(u')\}, \tag{9}$$

where f is obtained through replacing $A_{i,j}$ in A_i by $\mu_{i,j}(u_{i,j})$ and replacing "*and*", "*or*", "*not*" in A_i by "*min*", "*max*", "$1 - \mu$", respectively. And the output of all rules $A_1 \rightarrow B_1, \ldots, A_n \rightarrow B_n$, is fuzzy set M, which is defined as:

$$\forall u' \in U', \mu_M(u') = \max\{\mu_1(u'), \ldots, \mu_n(u')\}. \tag{10}$$

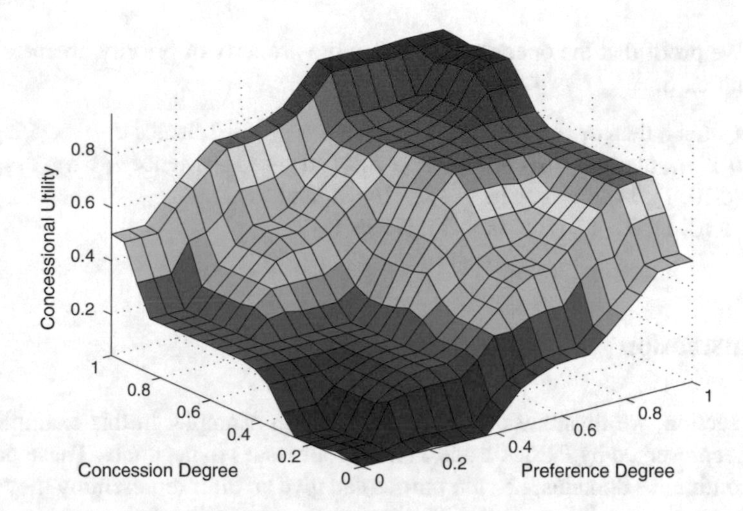

Fig. 4 The relation between preference degrees, concession degrees and concessional utilities. It shows that when a negotiator makes more concessions on a more important issue, compared with the ideal state, his descends more utilities. On the contrary, less concessions on a less important issue result in less concessional utilities

The result is still a fuzzy set and we should transform the fuzzy output into crisp onc. Such process is called defuzzification. We apply thc well-known centroid method [7] in this paper:

Definition 13 The centroid point u_{cen} of fuzzy set M given by formula (10) is:

$$u_{cen} = \frac{\int_{U'} u' \mu_M(u')du'}{\int_{U'} \mu_M(u')du'}. \tag{11}$$

Using this fuzzy inference method, we can obtain the relation between inputs (preference degrees, concession degrees) and output (concessional utilities), as shown in Fig. 4. After aggregating the above 20 fuzzy rules, our models can reflect humans' intuitive reasoning well.

4 A Priority Operator

In our model, when calculating the prioritised consistency degree of an offer, we need a priority operator, which should satisfy some properties shown in Definition 6. Here we apply a specific priority operator [6].

Theorem 1 *Operator* $\diamond$*:* $[0, 1] \times [0, 1] \to [0, 1]$*, which is defined as follows, is a priority operator:*

$$a_1 \diamond a_2 = a_1 \times (a_2 - 1) + 1 \tag{12}$$

Proof We proof that the operator $\diamond$ satisfies the property of priority operator listed in Definition 6.

(1) If $a_2 \leqslant a_2^{'}$, then $a_1 \times (a_2 - 1) + 1 \leqslant a_1 \times (a_2^{'} - 1) + 1$, hence $a_1 \diamond a_2 \leqslant a_1 \diamond a_2^{'}$,
(2) If $a_1 \leqslant a_1^{'}$, then $a_1 \times (a_2 - 1) + 1 \geqslant a_1^{'} \times (a_2 - 1) + 1$, hence $a_1 \diamond a_2 \geqslant a_1 \diamond a_2$,
(3) $\forall a \in [0, 1], 1 \diamond a = 1 \times (a - 1) + 1 = a$, and
(4) $\forall a \in [0, 1], 0 \diamond a = 0 \times (a - 1) + 1 = 1$ □

5 Illustration

In this section, we illustrate our model through an example. In this example, two parties, represented by Parties 1 and 2, have four issues to negotiate. These policies are in continuous domains, i.e., the parties can give an offer representing they partly supported or opposed the policies. Moreover, we assume that the acceptance degree can be accurate to 0.2 in this example, i.e., the acceptance domains can be divided into six levels from 0 to 1. The policy structures of both agents are summarised in Table 2. We use A, B, C, D to represent the following policies: *five percent tax increase*, *ten percent disarmament*, *ten percent increase of educational investment*, and *ten thousand economical housings investment*. Party 1 supports policies A and C, opposes policy D, and does not care about policy B (we call it irrelevant issue of Party 1, i.e., the outcome of policy B does not affect its utility), while Party 2 supports policy B, opposes policies A and C, and policy D is irrelevant for it.

In this example, we assume that two agents (i.e., acting on Parties 1 and 2) use a simple and friendly strategy in negotiation. It should be noted that an agent in a negotiation can use different negotiation strategies to achieve its goals. In this paper, our main work is to build an agent model that can represent well how concessions and preferences influence an outcome in real negotiation and find suitable evaluation methods of agreement. In order to illustrate how the evaluation methods work, we use a simple strategy, rather than complex ones. The strategy an agent uses in this example is as follows: (i) an agent gives the first offer showing friendliness, that is, using the opponent's attitude degree to calculate acceptance degree of irrelevant issues. For example, the first offer of Agent 1 is: $O_{i,1}(A) = 1$, $O_{i,1}(B) = 1$, $O_{i,1}(C) = 1$, and $O_{i,1}(D) = 0$. We denote this offer as $O_{i,1} = \{1, 1, 1, 0\}$ for short. Similarly, $O_{i,2} = \{0, 1, 0, 0\}$. (ii) An agent generates an offer according to concessional utility.

Table 2 Policy structure of agents

	A	B	C	D
Attitude of party 1	1	N/A	1	0
Preference of party 1	0.9	N/A	0.3	0.2
Attitude of party 2	0	1	0	N/A
Preference of party 2	0.2	0.5	0.7	N/A

The agent gives an offer with the least concessional utility first. If the opponent rejects this offer and the agent also rejects its opponent's counter-offer, the agent gives another offer with second least concessional utility in a new round. (iii) An agent accepts an opponent's offer if the concessional utility of the offer is not larger than that of its offer in the next step. That is, $\Delta\hat{u}_i(\lambda) \leqslant \Delta u_i^t(\lambda + 1)$ for the agent that starts the negotiation and $\Delta\hat{u}_i(\lambda) \leqslant \Delta u_i^t(\lambda)$ for the other one. We suppose Agent 1's discount factor is $\sigma_1 = 0.9$ and Agent 2's discount factor is $\sigma_2 = 0.8$. They use fuzzy reasoning method shown in Sect. 3 to calculate their concessional utilities.

The process of negotiation is shown in Table 3. In the first round, Agent 1 proposes an offer $(1, 1, 1, 0)$, then Agent 2 evaluates the offer through fuzzy reasoning and finds that it is higher than its acceptable threshold of concessional utility in this round $(\Delta u_2^t = 0.5251 > \Delta\hat{u}_2 = 0)$. Then Agent 2 gives an offer $(0, 1, 0, 0)$. Agent 1 also evaluates the offer and rejects it, and then it generates a new offer in the next round. After several rounds, Agent 2 accepts Agent 1's offer $(1, 1, 0, 0)$ finally. According to Definition 9, $(1, 1, 0, 0)$ is a utility-based agreement.

Similarly, we find a consistency-based agreement through consistency-based evaluation method. The strategy is based on prioritised consistency degree. More specifically, (i) an agent gives its first offer by using the opponent's attitude degree to calculate acceptance degree of irrelevant issues. (ii) An agent generates an offer according to prioritised consistency degree. The agent gives an offer with the highest prioritised consistency degree first. If there is no agreement in the first round, an offer with the second highest prioritised consistency degree will be proposed in a new round. (iii) An agent accepts an opponent's offer if the prioritised consistency degree of the offer is not smaller than that of its offer in the next step. This process of negotiation is shown in Table 4.

6 Experiment

This section demonstrates two experiments. The first experiment is proposed to reveal how the utility-based and consistency-based strategies influence the outcome of negotiations with different divisions of the acceptance intervals. The more divisions means that an agent can give an more accurate offer. The second experiment is proposed to analyse how the qualities of agreements can be improved with our utility-based method by comparing with an existing model [12].

6.1 *Experimental Setting*

In the first experiment, we set a negotiation scenario as follows: (i) two agents have two policies to negotiate and they have opposed attitudes to the policies; (ii) both agents have discount factor $\sigma = 0.9$; (iii) both agents use the same strategies shown in Sect. 5, including the utility-based and the consistency-based strategies; (iv) the

Table 3 The process of negotiation in the example based on concessional utility strategy

Round		A	B	C	D	Δu_2^t	$\Delta \hat{u}_2$	Respond		A	B	C	D	Δu_1^t	$\Delta \hat{u}_1$	Respond
1	Agent 1	1	1	1	0	0.5251	0	×	Agent 2	0	1	0	0	0.8655	0.1539	×
2	Agent 1	1	1	0.8	0	0.5348	0.1169	×	Agent 2	0.2	1	0	0	0.8780	0.1856	×
3	Agent 1	1	1	0.6	0	0.6061	0.1878	×	Agent 2	0.4	1	0	0	0.8207	0.2737	×
4	Agent 1	1	1	0.4	0	0.6475	0.3337	×	Agent 2	0.6	1	0	0	0.8280	0.4027	×
5	Agent 1	1	1	0.2	0	0.6684	0.4253	×	Agent 2	0.8	1	0	0	0.8438	0.5604	×
6	Agent 1	0.8	1	1	0	0.7652	0.5559	×	Agent 2	0	1	0.2	0	0.8439	0.5938	×
7	Agent 1	1	1	0	0	0.6620	0.6620	√	Agent 2							

Table 4 The process of negotiation in the example based on prioritised consistency strategy

Round		A	B	C	D	ρ_2	$\hat{\rho}_2$	Respond		A	B	C	D	ρ_1	$\hat{\rho}_1$	Respond
1	Agent 1	1	1	1	0	0.8367	1	×	Agent 2	0	1	0	0	0.7746	0.9899	×
2	Agent 1	1	1	0.8	0	0.8641	0.9933	×	Agent 2	0.2	1	0	0	0.8124	0.9798	×
3	Agent 1	1	1	0.6	0	0.8907	0.9866	×	Agent 2	0.4	1	0	0	0.8485	0.9695	×
4	Agent 1	0.8	1	1	0	0.8446	0.9798	×	Agent 2	0.6	1	0	0	0.8832	0.9695	×
5	Agent 1	1	1	0.4	0	0.9165	0.9764	×	Agent 2	0	1	0.2	0	0.7874	0.9592	×
6	Agent 1	0.8	1	0.8	0	0.8718	0.9730	×	Agent 2	0.8	1	0	0	0.9165	0.9592	×
7	Agent 1	1	1	0.2	0	0.9416	0.9695	×	Agent 2	0.2	1	0.2	0	0.8246	0.9487	×
8	Agent 1	0.8	1	0.6	0	0.8981	0.9661	×	Agent 2	1	1	0	0	0.9487	0.9487	√

acceptance intervals are divided into different levels to present how accurately an offer can be generated; (v) a policy's preference degree of each agent is randomly selected; and (vi) the experiments were repeated for one thousand times in each setting. In the second experiment, we compare our model with an axiomatic negotiation based model without an evaluation method of an offer. In that traditional model [12], agents negotiate through a minimal simultaneous concession, i.e., agents give up the least important policies in every round simultaneously until the remaining policies are not conflictive. We set a negotiation scenario as follows: (i) two agents have 10 indivisible policies (i.e., a policy should be totally supported or opposed) to negotiate, and the agents have different attitudes to n of them, where n will change from 1 to 10; (ii) both agents have discount factor $\sigma = 1$; (iii) a policy's preference degree of each agent is randomly selected; (iv) both agents use the utility-based strategy shown in the example and the minimal simultaneous concession solution; and (v) the experiments were repeated for one thousand times in each setting.

6.2 Results and Analysis

The first experimental results are showed in Figs. 5 and 6. In Fig. 5, the 'dot' line is used to show how the average concessional utilities of both agents in an agreement change when the number of division of acceptance interval increases, while the 'cross' line represents the best situation where both agents obtain the lowest average concessional utilities in the situation where they collaborate with each other. Comparing these two lines, we can see clearly that the average concessional utility of both agents increases when the number of division increases, while the minimum values are similar. Figure 6 shows how the prioritised consistency degree in agreement changes through consistency-based strategy when the number of division of acceptance interval increases. The 'dot' line presents an average prioritised consis-

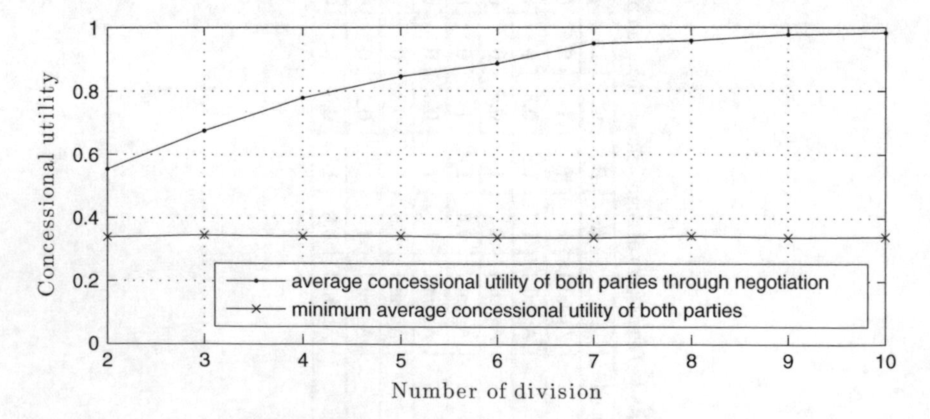

Fig. 5 Relation between number of division and concessional utilities

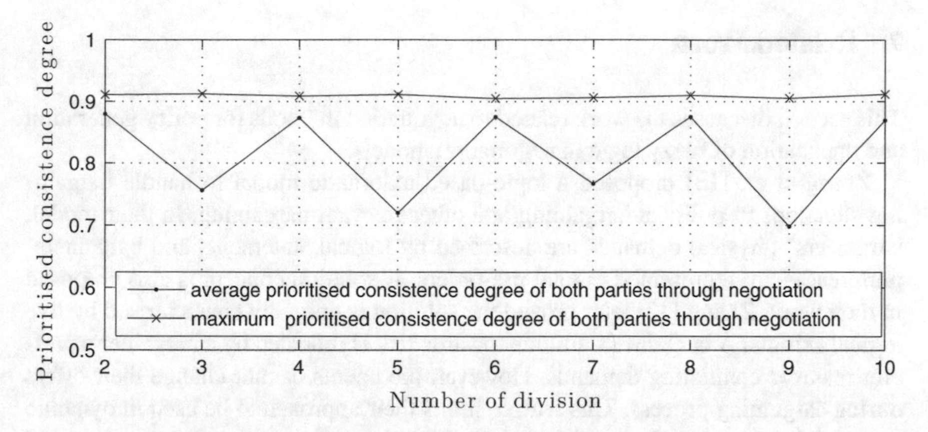

Fig. 6 Relation between number of division and prioritised consistency

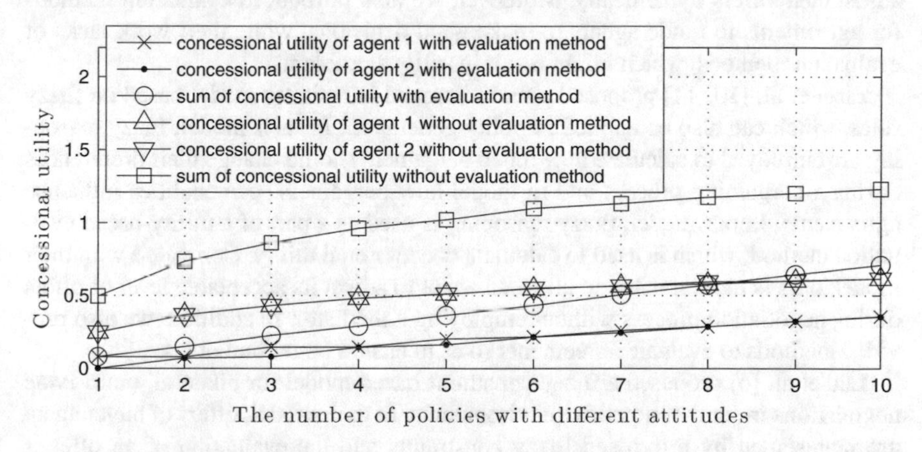

Fig. 7 Concessional utilities in agreement with and without evaluation method

tency degree of both agents, while the 'cross' line represents the best situation where both agents obtain the highest average prioritised consistency degree. Comparing these two lines, we can see that the average prioritised consistency degree does not change too much (just between 0.7 and 0.9), and is not far away from the highest prioritised consistency degree.

The results in the second experiment are showed in Fig. 7. The figure reveals that under the guide of evaluation method based on concessional utility, our model could avoid more losses than the one of [12]. Comparing the line marked by squares and the line marked by circles, we can see that using an evaluation method, both agent can save almost a half loss together, even when the number of policies with different attitudes increases.

7 Related Work

This section discusses the work related to negotiation methods for policy generation and application of fuzzy logic in negotiation models.

Zhang et al. [13] proposed a logic-based axiomatic model to handle bargaining situations in political bargaining and other relevant bargaining. In their model, bargainers' physical demands are described by logical statements and bargainers' preferences are represented in total pre-orders. A solution concept is also proposed in their work. Zhang [12] later proved the solution is uniquely characterised by five logical axioms. A bargainer's attitude towards risk is reflected by a bargainer's preferences over conflicting demands. However, the agents cannot change their offers during bargaining process. This feature limits their approach to be used in dynamic and complex domains. Our model used an alternating-offers protocol, so agents could adjust their offers dynamically. Moreover, we also introduced evaluation methods for agreements to guide agents to make good decisions, while their work lacks of evaluation methods, which might result in serious problems.

Zhan et al. [10, 11] proposed a multi-demand bargaining model based on fuzzy rules, which can also be applied in polity generation. In their model, fuzzy reasoning are employed to calculate how much bargainers should change their preferences during a bargaining process and to model how bargainers' personalities influence agreements. In our model, fuzzy reasoning is used as a part of a utility-based evaluation method, which is used to calculate concessional utility. Compared with their model, ours is more flexible to allow an agent to adjust its acceptable level of offers during negotiation process without employing a mediator. In addition, we also provided methods to evaluate agreements so as to ensure better outcomes.

Luo et al. [6] proposed a fuzzy constraint-based model for bilateral, multi-issue negotiations in semi-competitive environments. In their model, offers of buyer agent are represented by prioritised fuzzy constraints and the evaluation of an offer is regards as a prioritised fuzzy constraint satisfaction problem. During a negotiation, a buyer agent sends its fuzzy constraints according to priority. If the seller agent cannot give an offer under the constraints given by the buyer, the buyer should relax some of its constraints. However, in each round of their negotiation a buyer agent can only submit one new constraint or relax a submitted one. This feature might limit their approach to be used in complex domains. In our model, agents can change its offer flexibly and dynamically under the guide of two evaluation methods of offers.

8 Conclusions and Future Work

In this paper, a negotiation-based policy generation model was proposed to solve the problem of lacking of agreement evaluation methods in existing work and guide agents to make appropriate decisions to achieve better outcomes. The first one is a utility-based evaluation method. It handles reasoning about how concessions and preferences of policies affect an agent's utility. In this method, a fuzzy reasoning was used to calculate concessional utility of an offer for an agent based on intuitive fuzzy rules. The second one is a consistency-based evaluation method. It calculates the similarity between an agent's ideal state and an offer. The similarity is represented by a prioritised consistency degree, and a specific priority operator is used to express the influence of preferences of policies contributing to consistency. The experimental results showed that different evaluation methods of offers can result in different outcomes and the proposed policy generation model can efficiently handle situations where utilities of policies are hard to obtain, and successfully lead the political negotiation to an agreement. With the help of evaluation methods, an agent can make less utility concessions than a traditional one.

The future work will pay attention to negotiation strategies. An extended negotiation protocol will also be developed by adding an information transfer module between agents or in a coalition to model the cooperation between the parties in policy generation scenarios.

Acknowledgements This work is supported by an International Program for Ph.D. Candidates from Sun Yat-Sen University and an IPTA scholarship from University of Wollongong.

References

1. K.J. Arrow. *Social Choice and Individual Values*, vol. 12 (Yale University Press, New Haven, 2012)
2. W.D. Cook, Distance-based and ad hoc consensus models in ordinal preference ranking. Eur. J. Oper. Res. **172**(2), 369–385 (2006)
3. R. Davis, R.G. Smith, Negotiation as a metaphor for distributed problem solving. Artif. Intell. **20**(1), 63–109 (1983)
4. P.C. Fishburn, S.J. Brams, Paradoxes of preferential voting. Math. Mag. 207–214 (1983)
5. X. Jing, D. Zhang, X. Luo, A logical framework of bargaining with integrity constraints, in *AI 2013: Advances in Artificial Intelligence* (Springer, Heidelberg, 2013), pp. 1–13
6. X. Luo, N.R. Jennings, N. Shadbolt, H.-F. Leung, J.H.-M. Lee, A fuzzy constraint based model for bilateral, multi-issue negotiations in semi-competitive environments. Artif. Intell. **148**(1), 53–102 (2003)
7. E.H. Mamdani, S. Assilian, An experiment in linguistic synthesis with a fuzzy logic controller. Int. J. Man Mach. Stud. **7**(1), 1–13 (1975)
8. A. Piegat, *Fuzzy Modeling and Control*, Studies in Fuzziness and Soft Computing (Physica-Verlag HD, 2010)
9. A. Rubinstein, Perfect equilibrium in a bargaining model. Econom. J. Econom. Soc. **50**(1), 97–109 (1982)
10. J. Zhan, X. Luo, K.M. Sim, C. Feng, Y. Zhang, A fuzzy logic based model of a bargaining game, in *Knowledge Science, Engineering and Management*, Lecture Notes in Artificial Intelligence, vol. 8041 (Springer, Heidelberg, 2013), pp. 387–403
11. J. Zhan, X. Luo, C. Feng, W. Ma, A fuzzy logic based bargaining model in discrete domains: Axiom, elicitation and property, in *2014 IEEE International Conference on Fuzzy Systems* (2014), pp. 424–431
12. D. Zhang, A logic-based axiomatic model of bargaining. Artif. Intell. **174**(16), 1307–1322 (2010)
13. D. Zhang, Y. Zhang, An ordinal bargaining solution with fixed-point property. J. Artif. Intell. Res. **33**(1), 433–464 (2008)

Fixed-Price Tariff Generation Using Reinforcement Learning

Jonathan Serrano Cuevas, Ansel Y. Rodriguez-Gonzalez and Enrique Munoz de Cote

Abstract Tariff design is one of the fundamental building blocks behind distributed energy grids. Designing tariffs involve considering customer preferences, supply and demand volumes and other competing tariffs. This paper proposes a broker capable of understanding the market supply and demand constraints to issue time-independent tariffs that can be offered to customers (energy producers and consumers) on smart grid tariff markets. The focus of this work is laid on determining the most profitable price on time-independent tariffs. While this type of tariffs are the most simple of all, it allows us to study the fundamental underpinnings behind determining tariff prices considering imperfect and semi-rational customers and competing tariffs. Our proposed broker agent —COLD Energy— learns its opponents strategy dynamics by reinforcement learning. However, as opposed to similar methods, its advantage lies in its ability to learn fast and adapt to changing circumstances by using a sufficient and compact representation of its environment. We validate the proposed broker in Power TAC, an annual international trading agent competition that gathers experts from different fields and latitudes. Our results show that the proposed representation is capable of coding the important characteristics of tariff energy markets for fixing energy prices when the competing brokers are non-stationary (learning), irrational, fixed, rational or greedy.

1 Introduction

Together with the adoption of smarter energy grids comes the idea of deregulating the energy supply and demand through energy markets, where producers are able to sell

J. Serrano Cuevas (✉) · A.Y. Rodriguez-Gonzalez · E. Munoz de Cote
Instituto Nacional de Astrofísica, Óptica y Electrónica, Puebla, Mexico
e-mail: jonathan.serrano@ccc.inaoep.mx

A.Y. Rodriguez-Gonzalez
e-mail: ansel@ccc.inaoep.mx

E. Munoz de Cote
e-mail: jemc@inaoep.mx

© Springer International Publishing AG 2017
K. Fujita et al. (eds.), *Modern Approaches to Agent-based Complex Automated Negotiation*, Studies in Computational Intelligence 674,
DOI 10.1007/978-3-319-51563-2_8

energy to consumers by using a *broker* as an intermediary. One of the most dominant energy markets is the tariff market, where small consumers can buy energy from broker agents[1] via tariffs. Tariffs are contracts agreed between either a producer or a consumer, and a broker, which entitle both parts the right to trade a certain amount under certain conditions [1]. These conditions might include the payment per amount of energy traded, minimum signup time, signup or early withdraw payments, among others [2]. It is through an open energy market of this kind, which uses tariffs to buy and sell energy that the gross majority of the traded energy takes place. For this reason, this work is focused on proposing a tariff-expert broker agent for the tariff energy markets. We use Power TAC [3], an annual international trading agent competition that gathers experts from different fields and latitudes to validate our proposed broker. Power TAC is a complex simulator of an entire energy grid with producers, consumers and brokers buying and supplying energy. It considers transmission and distribution costs, models many different types of energy generation and storage capacities and uses real climate conditions and user preferences to simulate the environment where brokers should take autonomous decisions.

Several aspects, including the customers' preferences and the competitions' offers, were taken into account to design our tariff-expert broker [4], which uses reinforcement learning to generate electric energy tariffs while striving to maximize its utility on the long term. To test our proposed tariff design, we embedded our solution in COLD Energy, a broker agent that considers many other aspects of the smart-grid (like a wholesale day-ahead and spot markets, balancing issues and portfolio management). However, this paper will focus solely on the tariff maker part of COLD Energy.

The paper is structured as follows, in Sect. 2 we present a general background on Power TAC and the electricity tariff markets. Then we present the most relevant work related to ours. In Sect. 3 we present our tariff-expert contribution embedded in COLD Energy. We present our experimental results in Sect. 4 and close our work with some relevant conclusions.

2 Power TAC and Tariff Markets

Power TAC [3] is a smart grid [5] simulation platform where a set of brokers compete against each other in an energy market. Power TAC uses a multi-agent approach [6] to simulate a smart grid market, where brokers can buy or sell energy to their customers in two different markets: the wholesale market and the tariff market, however, this paper is focused solely on the tariff market. In the tariff market, the brokers trade energy with their clients by using contracts called tariffs, which include specifications such as price-per-kwh, subscription or early withdrawal fees, periodic payments and, the most important one: price. The experiments on this paper used a particular type of tariff called flat tariff [7]. A flat tariff is a time independent tariff, which offers

[1]Note that we refer to brokers and agents indistinctly.

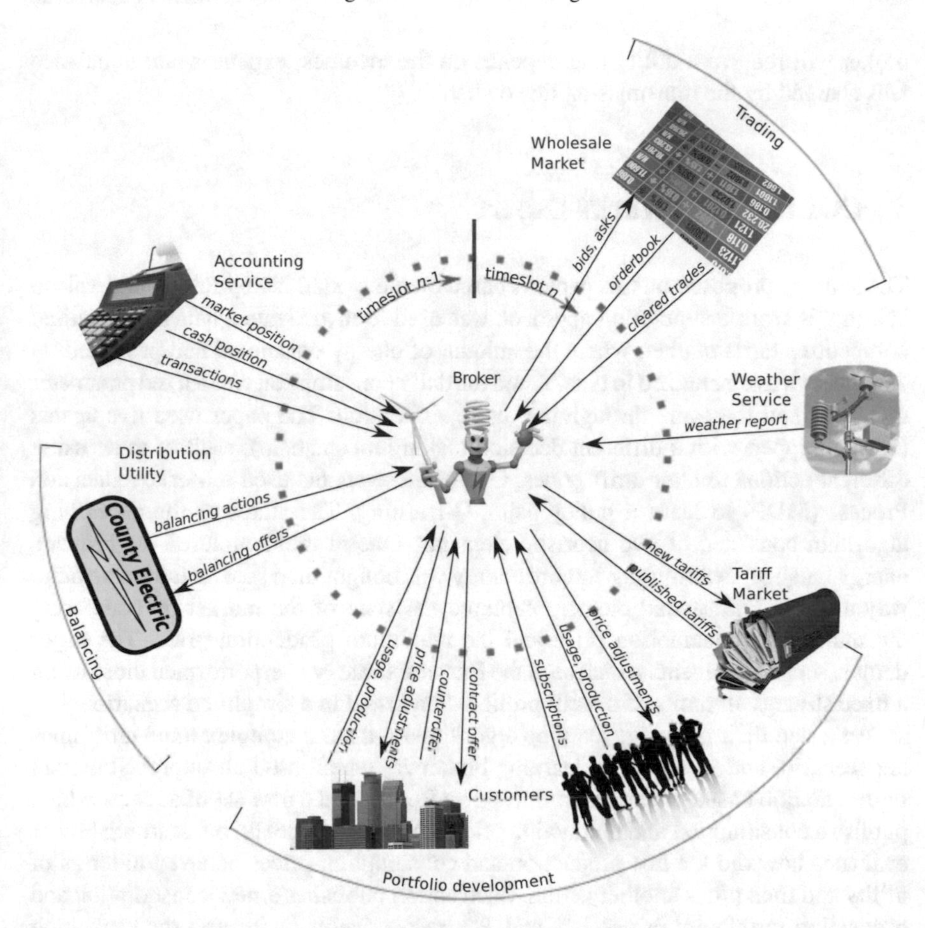

Fig. 1 PowerTAC timeslot cycle including the tariff market operations

a fixed price per energy unit disregarding the time, i.e. the time of the day of the day of the week; therefore its only specification is price-per-kwh. Figure 1 shows the Power TAC cycle, including the tariff market period. During this period each broker publishes tariffs, and customers evaluate them and decide if they should subscribe to them. Later on this period the consumption and production operations related to tariffs are executed, and the transaction proceedings are charged either to the brokers or to the customers at the end of each time unit. The time unit used on Power TAC is a timeslot, which represent one simulated hour. The brokers can publish tariffs at any given timeslot. After publishing a tariff, the customers can evaluate the offers and decide if they are to stay with the same tariff or change to any available tariff, which may belong to the same broker or to another one. The objective of every single broker is to publish attractive tariffs, so that the producing-customers want to sell energy to it and the consuming-customers want to buy energy from it. At the end, every broker will receive a utility that depends on the incomes, expenses and unbalance fees charged by the transmission line owner.

3 COLD Energy Tariff-Expert

The strategy proposed on this paper is based on the work done by Reddy and Veloso [8]. In this work a simulation approach was used to investigate a heavily simplified competitive tariff market, where the amount of energy consumed and produced by customers was discretized in blocks, and the daily consumption was a fixed parameter that remained the same through the entire simulation. The paper used five agents (each equipped with a different decision making mechanism), each of them using different actions to alter tariff prices. One of these agents used a Markov Decision Process (MDP) to learn a policy using Q-learning. The states of the Q-learning algorithm consisted of two heuristic elements. One of them captured the broker's energy balance, determining if more energy was bought than sold or it was the other way around. The second element captured the state of the market by comparing the minimum consumption price and the maximum production price. The paper demonstrated that agents which used the learning strategy overperformed those using a fixed strategy in terms of overall profit, when tested in a simplified scenario.

We tested their proposed learning algorithm on a more complex fixed-tariff market scenario, and developed a learning broker B_L which used an improved market representation based on the one proposed by Reddy, and a new set of actions, which publish a consumption and production price each. In more detail, our learning broker evaluates how did the last production and consumption prices behaved in terms of utility and then picks another action. Each action publishes a new consumption and production tariff with prices $P^{B_k}_{t,C}$ and $P^{B_k}_{t,P}$ respectively. At the end the evaluation period, $\Psi_{t,C}$ and $\Psi_{t,P}$ represent the amount of energy sold or acquired by the broker respectively. In general terms the literal P will be used to refer to an energy price and Ψ to refer to an energy amount. For each evaluation period, the utility function for broker k (B_k) is the one shown in Eq. 1. The first term represents the income total proceedings due to electric energy sale, the second terms corresponds to the amount paid to producers, and the third term represents an inbalance fee.

$$u^{B_k}_t = P^{B_k}_{t,C}\Psi_{t,C} - P^{B_k}_{t,P}\Psi_{t,P} - \theta_t|\Psi_{t,C} - \Psi_{t,P}| \tag{1}$$

Each term in Eq. 1 represents either a monetary income or outcome. So the whole utility represents a monetary amount. All three terms multiply a price per energy unit by an energy amount, yielding a monetary unit. If the difference $\Psi_{t,C} - \Psi_{t,P}$ equals zero, then the broker sold exactly the same amount of energy it bought, so the energy inbalance is zero; and for this reason the inbalance fee is zero as well. The variable θ_t is the amount the broker has to pay to the transmission line owner per each unit of energy inbalance it generated on the evaluation period.

The utility function from Eq. 1 was used as the MDP's reward after executing a certain action at a given state while in time t. Our brokers state representation will be described on Sect. 3.3 and its actions on Sect. 3.4.

3.1 Market Model

It is important to mention in first place that the market model was designed with the purpose of being used to maximize the utility in the long term. The environment description, encoded as discrete states depend on some key elements belonging to the tariffs published by other brokers; namely: maximum and minimum consumption prices, and maximum and minimum production prices. These parameters are described in the following way.

Minimum consumption price:

$$P_{t,C}^{min} = min_{B_k \in B \setminus \{B_L\}} P_{t,C}^{B_k} \tag{2}$$

Maximum consumption price:

$$P_{t,C}^{max} = max_{B_k \in B \setminus \{B_L\}} P_{t,C}^{B_k} \tag{3}$$

Minimum production price:

$$P_{t,P}^{min} = min_{B_k \in B \setminus \{B_L\}} P_{t,P}^{B_k}, \tag{4}$$

Maximum production price:

$$P_{t,P}^{max} = max_{B_k \in B \setminus \{B_L\}} P_{t,P}^{B_k}, \tag{5}$$

where B_L represents the learning broker evaluating these parameters and the minimum and maximum prices are taken from a list conformed by the prices of all the other brokers, but not the prices of the learning broker B_L. Now we will proceed to explain the MDP we used.

3.2 MDP Description

The MDP used by COLD Energy is shown in Eq. 6.

$$M^{B_L} = \langle S, A, P, R \rangle \tag{6}$$

where:

- $S = \{s_i : i = 1, \ldots, I\}$ is a set of I states,
- $A = \{a_j : j = 1, \ldots, J\}$ is a set of J actions,
- $P(s, a) \rightarrow s\prime$ is a transition function and
- $R(s, a)$ equals $u_t^{B_{k=L}}$ and represents the reward obtained for execution action a while in state s.

3.3 States

A series of states were designed so as to provide our learning broker of a discretized version of the market, which considers as well the effect of the actions executed by the other brokers. Specifically the state space S is the set defined by the following tuple:

$$S = \langle PRS_t, PS_t, CPS_t, PPS_t \rangle \tag{7}$$

where:

- $PRS_t = \{rational, inverted\}$ is the price range status at time t and
- $PS_t = \{shortsupply, balanced, oversupply\}$ is the portfolio status at time t.
- $CPS_t = \{out, near, far, veryfar\}$ is the consumers price status,
- $PPS_t = \{out, near, far, veryfar\}$ is the producers price status,

The values PRS_t and PS_t capture the relationship between the highest production price and the lowest consumption price, and the balance of the broker B_L, respectively. This two parameters were proposed by Reddy and are defined as follows:

$$PRS_t = \begin{cases} rational \text{ if } P_{t,C}^{min} > P_{t,P}^{max} \\ inverted \text{ if } P_{t,C}^{min} \leq P_{t,P}^{max} \end{cases} \tag{8}$$

$$PS_t = \begin{cases} balanced \quad \text{if } \Psi_{t,C} = \Psi_{t,P} \\ shortsupply \text{ if } \Psi_{t,C} > \Psi_{t,P} \\ oversupply \quad \text{if } \Psi_{t,C} < \Psi_{t,P} \end{cases} \tag{9}$$

where:

- $P_{t,C}^{min} = min_{B_k \in B \setminus \{B_L\}} P_{t,C}^{B_k}$ is the minimum consumption price,
- $P_{t,C}^{max} = max_{B_k \in B \setminus \{B_L\}} P_{t,C}^{B_k}$ is the maximum consumption price,
- $P_{t,P}^{min} = min_{B_k \in B \setminus \{B_L\}} P_{t,P}^{B_k}$ is the minimum production price and
- $P_{t,P}^{max} = max_{B_k \in B \setminus \{B_L\}} P_{t,P}^{B_k}$ is the maximum production price

On these equations B_L represents the learning broker evaluating these parameters. So the minimum and maximum prices consider the list conformed by the prices of all the other brokes but not the prices of the learning broker B_L.

These two elements of S encode the price actions of the broker related to the prices of the other brokers. These parameters, as coarse as they can be, create a compact representation of a market that might include several brokers publishing many tariffs. This representation's size will remain unchanged disregarding the latter factors, but at the same time the representation will capture the tariff market price states as a whole, considering the other competing brokers' tariff publications. The tuple parameters CPS_t and PPS_t can take any of these values: *out, close, far, very far* and are defined as follows.

$$CPS_t = \begin{cases} out & \text{if } Top_{ref} \leq P^{B_L}_{t-1,C} \\ near & \text{if } Thres_{ref} < P^{B_L}_{t-1,C} \leq Top_{ref} \\ far & \text{if } Middle_{ref} < P^{B_L}_{t-1,C} \leq Thres_{ref} \\ veryfar & \text{if } P^{B_L}_{t-1,C} \leq Middle_{ref} \end{cases} \tag{10}$$

where:

- $Top_{ref} = P^{min}_{t,C}$
- $Middle_{ref} = \frac{P^{min}_{t,C} + P^{min}_{t,P}}{2}$
- $Thres_{ref} = \frac{Top_{ref} + Middle_{ref}}{2}$

$$PPS_t = \begin{cases} out & \text{if } Bottom_{ref} \geq P^{B_L}_{t-1,P} \\ near & \text{if } Thres_{ref} \geq P^{B_L}_{t-1,P} > Bottom_{ref} \\ far & \text{if } Middle_{ref} \geq P^{B_L}_{t-1,P} > Thres_{ref} \\ veryfar & \text{if } P^{B_L}_{t-1,P} \geq Middle_{ref} \end{cases} \tag{11}$$

where:

- $Bottom_{ref} = P^{min}_{t,P}$,
- $Middle_{ref} = \frac{P^{min}_{t,C} + P^{min}_{t,P}}{2}$
- $Thres_{ref} = \frac{Bottom_{ref} + Middle_{ref}}{2}$

3.4 Actions

The set of actions is defined as:

$$A = \{maintain, lower, raise, inline, revert, minmax, wide, bottom\} \tag{12}$$

Each one of these actions define how the learning agent B_L determines the prices $P^{B_L}_{t+1,C}$ and $P^{B_L}_{t+1,P}$ for the next timeslot t+1. These actions were designed so as to provide the broker with several ways to react fast to market changes. It is important to recall that every single action impacts both the production and consumption price features of the next tariffs to be published. These are the specific details of each action:

- *maintain* publishes the same price as in timeslot t−1.
- *lower* decreases both consumer and producer prices by a fixed amount.

- *raise* increases both the consumer and producer prices by a fixed amount.
- *inline* sets the consumption and production prices as $P^{B_L}_{t+1,C} = \lceil m_p + \frac{\mu}{2} \rceil$ and $P^{B_L}_{t+1,P} = \lfloor m_p - \frac{\mu}{2} \rfloor$.
- *revert* moves the consumption and production prices towards the midpoint $m_p = \lfloor \frac{1}{2}(P^{min}_{t,C} + P^{min}_{t,P}) \rfloor$.
- *minmax* sets the consumption and production prices as $P^{B_L}_{t+1,C} = D_{coeff} P^{max}_{t,C}$ and $P^{B_L}_{t+1,P} = P^{min}_{t,P}$, where D_{coeff} is a number on the interval [0.70, 1.00] which damps the effect of the minmax action over the consumption price.
- *wide* increases the consumption price by a fixed amount ε and decreases the production price by a fixed amount ε.
- *bottom* sets the consumption price as $P^{B_L}_{t+1,C} = P^{min}_{t,C} \dot{M}argin$, where the production price $P^{B_L}_{t+1,P} = P^{min}_{t,P}$. The Bottom action is market-bounded.

3.5 State/Action Flow Example

To illustrate an action's effect over the consumption and production prices, Fig. 2 shows a simple simulated flow on a series of actions. The actions appear above the graph. On this hand-made simple scenario COLD Energy competes against two brokers, who publish one consumption and one production tariff each. The horizontal axis represents the time measured in decision steps, the vertical axis corresponds to the energy price. The dashed lines are fixed references, while the continuous lines are the published prices as described below:

- maxCons: corresponds to $P^{max}_{t,C}$ and is equal to 0.5. It can be assumed that competing broker A published a consumption tariff with this price.

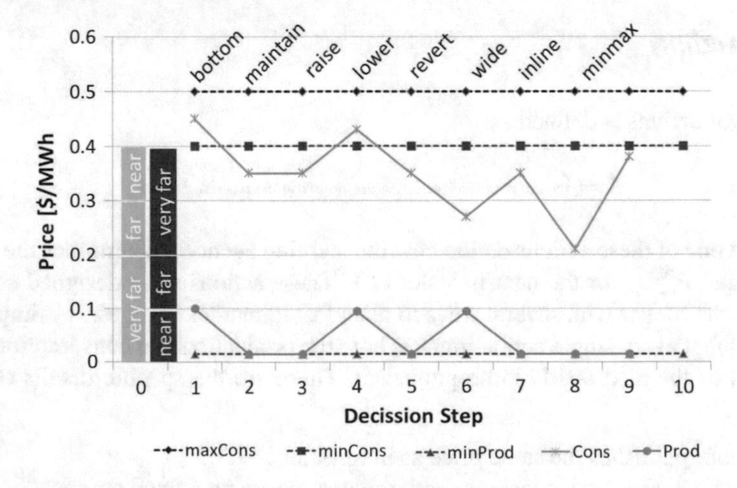

Fig. 2 Overall average and standard deviation for each broker

- minCons: corresponds to $P_{t,C}^{min}$ and is equal to 0.4. It can be assumed that competing broker B published a consumption tariff with this price.
- minProd: corresponds to $P_{t,P}^{min}$ and $P_{t,P}^{max}$; which means that the maximum and minimum production prices are the same and is equal to 0.015. It can be assumed that both brokers A and B published a production tariff with this price.
- Cons: corresponds to the consumption price published by COLD Energy.
- Prod: corresponds to the production price published by COLD Energy.

COLD Energy will bound the price range of its tariffs in the range $[P_{t,P}^{max}, P_{t,C}^{min}]$. For this reason, none of the actions will lead to a price position outside this range. This feature ensures that any consumption price published by Cold Energy will be more attractive to energy buyers, and any production price published will be more attractive to energy sellers.

The learning algorithm used was the Watkins-Dayan [9] Q-Learning update rule with an $\varepsilon - Greedy$ exploration strategy. This strategy either selects a random action with ε probability or selects an action with $1 - \varepsilon$ probability that gives maximum reward in a given state.

$$\hat{Q}_t(s,a) \leftarrow (1-\alpha_t)\hat{Q}_{t-1}(s,a) + \alpha_t \left[r_t + \gamma \hat{Q}_{t-1}(s',a') \right], \quad (13)$$

a'

4 Experimental Results

This section will describe the results obtained by using the market representation and the actions described on the previous section. Six different brokers participated on the series of experiments, including COLD Energy and ReddyLearning. The different brokers are described on Table 1.

Table 1 Competing brokers general description

COLD Energy	The learning broker developed on this thesis work
ReddyLearning	The learning broker proposed by Reddy
Fixed	Publishes a initial production and consumption tariff and never updates them again
Balanced	A fixed-strategy broker which uses the Balanced strategy proposed by Reddy
Greedy	A fixed-strategy broker which uses the Greedy strategy proposed by Reddy
Random	A broker that uses COLD Energy's market representation and actions. This broker chooses randomly among the available actions at each evaluation period

Since COLD Energy deals with flat tariffs, it is necessary to test our broker against similar ones. For this reason the broker ReddyLearning was chosen. The same logic applies for the selection of the remaining brokers. It is not possible to tell the result of the pricing strategy apart if the tariff creation mechanisms of the competing and if the competing brokers are not publishing only flat tariffs. These two considerations are really important since Power TAC provides the capability of publishing time-dependent tariffs and also supplies wholesale market abilities to every broker.

4.1 General Setup

Prior to the experiments, both COLD Energy and ReddyLearning were trained against a fixed broker for 2,000 timeslots and against the random broker for 8,000 timeslots. During the training sessions the brokers were adjusted to explore at every decision step, updating their Q-table with the obtained reward. The trained Q-table was stored and transferred to the brokers to be exploited on the experiments. The experimental general setup includes a game length of 3000 timeslots and a tariff publication interval of 50 $\pm$ 5 timeslots when a consumption and a production tariffs are published. Lastly, since the training process took place already before the experimental session, the learning brokers did not explore at all during the test sessions.

4.2 Experiments Description

The experiments were designed to test COLD Energy against specific sets of the competing brokers and itself. We conducted the following set of experiments.

- COLD Energy versus All: our learning broker versus Random, Balanced, Greedy and the learning broker proposed by Reddy, named as ReddyLearning.
- COLD Energy versus ReddyLearning: our learning broker versus the learning broker proposed by Reddy.

4.3 COLD Energy Versus All

This series of experiments included all the brokers. Figure 3 plots the average and standard deviation per publication interval for each broker, while Fig. 4 is an example of how the accumulated utility behaved on one of the experiments.

Several observations can be drawn from these results. First, Fig. 3 clearly show that COLD Energy has the highest utility compared to the rest of the competing brokers. The second position is for the Random broker and the third one for ReddyLearning. The latter broker uses the market representation and set of actions proposed by Reddy

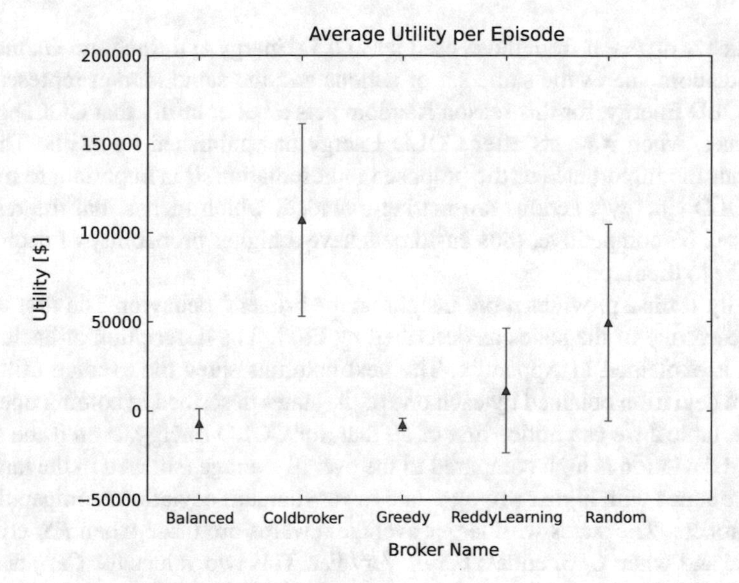

Fig. 3 Overall average and standard deviation for each broker

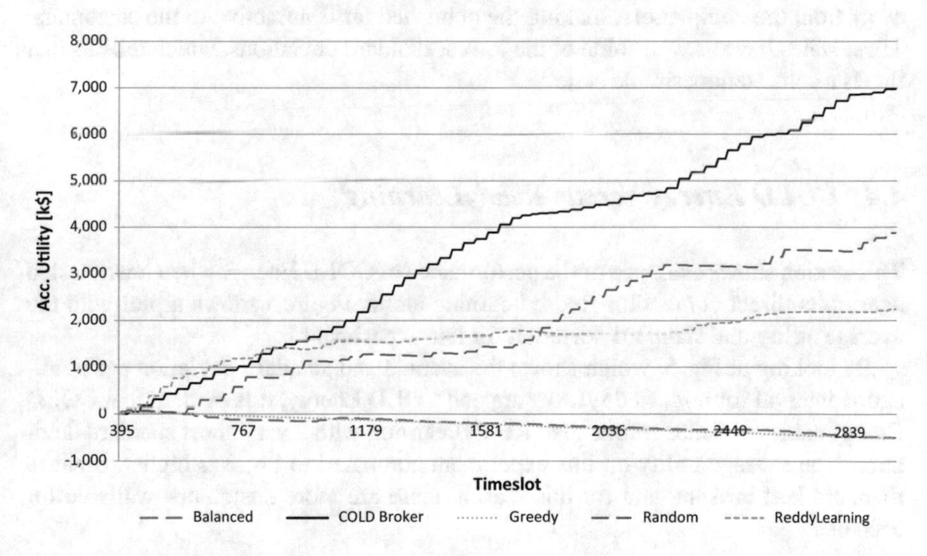

Fig. 4 Accumulated utility for each broker

[8], which is different from those used by COLD Energy and Random. On the other hand, Random shares the same set of actions and the same market representation with COLD Energy, for this reason Random gets a better utility that COLD Energy sometimes, when it reacts after COLD Energy has published its tariffs. This fact highlights the importance of the proposed representation. It is important to mention that COLD Energy's actions are market-bounded, which means that the resulting prices will be competitive, thus customers have a higher probability of deciding to subscribe to them.

Finally Table 2 provides more insight on the brokers' behavior. The first column shows each one of the states as described by Eq. 7. The description of each abbreviation is explained in Appendix. The next columns show the average utility and standard deviation obtained by each one of the states described in column one. If we observe Table 2 we can notice first of all that, for COLD Energy, even if the overall standard deviation is high compared to the overall average (showed in the last row), there are states with higher averages and lower standard deviations compared to the other brokers. The states with larger average rewards are those when PS_t equals to Rational and when CPS_t equals Far or Very Far. This two values for CPS_t are associated with the *inline* and *bottom* actions, which safely place the consumption price away from the competitors, making the published tariff attractive to the customers. These states have as well some of the lowest standard deviations, which tells us that this is a consistent desirable state.

4.4 COLD Energy Versus ReddyLearning

This section shows evidence of the performance of COLD Energy when it was tested against its direct competitor ReddyLearning alone. Figure 5 shows a plot with the average utility and standard deviation for this experiment.

By looking at Fig. 5, which shows the average and standard deviation per publication interval for both ReddyLearning and COLD Energy, it is evident that COLD Energy achieves better results than ReddyLearning with a very short standard deviation. The average utility on this experiment compared to Fig. 3 is higher, because there are less brokers, and for this reason, there are more customers available for each one.

5 Conclusions

The experiments showed that COLD Energy, with its proposed set of actions and its market representation was able to obtain the highest profits 70% of the evaluated timeslots when tested against all the competing brokers, including ReddyLearning. When tested only against the latter, COLD Energy was able to obtain the highest profit 100% of the evaluated timeslots. This proved that both the market representation and

Table 2 Average and standard deviation per state and broker

	Balanced		Cold		Greedy		ReddyLearning		Random	
State	Average	Std. Dev	Average	Std. Dev	Average	Std. Dev	Average	Std. Dev	Average	Std. Dev
RashFaOu			1,74,153	11,625					1,02,456	73,811
RashVeOu			1,73,011	18,419					1,30,540	
RashFaNe			1,70,911	20,262					72,790	1,14,024
RashNeOu			1,55,069	24,060					99,016	88,380
RashOuFa	261	1,48,156	16,380	– 7,513	–	1,47,667			– 6,428	4,495
RashNeNe			1,32,635	9,086						
RashOuOu			1,29,775	46,235	– 7,960	125			24,379	48,751
RashOuVe	– 6,313	2,311	1,26,248				23,663	37,111	10,078	62,824
InshNeNe			1,22,193	46,484					82,709	41,243
InshOuFa	– 5,511	4,473	1,11.042	21,857	– 7,786	308			30,420	48,639
InshOuVe	– 11,875	7,083	99,340	40,064	– 7,728	343	1,141	15,419	43,969	45,035
InshNeOu			98,531	41,451					79,228	45,536
InshFaNe			95,388	27,956					92,427	26,106
InshVeNe			91,703	–					1,06,631	7,481
InshOuNe	– 5,172	2,305	89,993	3,679	– 7,797	288			23,464	41,034
InshFaOu			86,998	53,118					90,494	39,285
InshVeOu			86,447	45,530					1,12,301	49,444
InshOuOu	– 4,757	3,428	56,708	28,661	– 7,809	258			19,789	35,326
RashFaVe	58,018		52,400		44,510		51,925		80,048	
RaovOuVe	– 15,966	18,831					6,454	17,590		

(continued)

Table 2 (continued)

	Balanced		Cold		Greedy		ReddyLearning		Random	
InovOuOu	− 2,213	8,171								
InshNeVe							1,45,151			
InovOuVe	− 8,364	4,235					1,762	15,281		
InovOuFa	− 4,015	4,253								
RashNeVe							1,61,590	53,707		
InshNeFa									56,759	56,130
RashOuNe									2,636	22,328
Summary	**− 7,008**	**8,317**	**1,07,238**	**54,109**	**− 7,552**	**3,491**	**11,459**	**34,963**	**49,507**	**55,078**

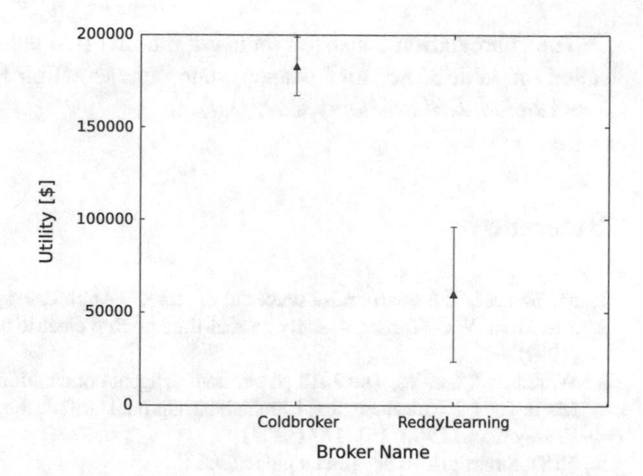

Fig. 5 Overall average and standard deviation for COLD Energy and ReddyLearning

the proposed actions achieved a better average utility compared to that delivered by the other competing brokers against whom it was tested, namely ReddyLearning, Balanced, Greedy and Random.

It is important to mention as well that the market representation size is not bounded to the number of competing brokers; the number of possible value combinations of state space S will remain the same if there are 1, 2 or more competing brokers. This is very useful because it makes easier the learning process. On the other hand, the market-bounded actions proposed were the most used by COLD Energy, and these actions conducted it to lead the utility rank most of the time on the experiments executed. Even as there were some non-market-bounded actions available, such as *Minmax* for instance, COLD Energy learned that those actions did not yield good results, and for this reason decided not to use them.

Appendix

In order to keep clean and reduced tables, some abbreviations were used to designate the names of the values each state can take.

Table 3 States values and abbreviations

PRS_t	Rational(Ra), Inverted(In)
PS_t	Shortsupply(sh), Balanced(ba), Oversupply(ov),
CPS_t	Very Far(Ve), Far(Fa), Near(Ne), Out(Ou)
PPS_t	Very Far (Ve), Far(Fa), Near(ne), Out(ou)

The abbreviation consisted on using the first two letters of the value's name, as stated on Table 3. So, for instance, state representation RaShFaOu stands for state $S = \langle Rational, Shortsupply, Far, Out \rangle$.

References

1. M. Wissner, The smart grid-a saucerful of secrets? Appl. Energy **88**(7), 2509–2518 (2011)
2. J.A. Mont, W.C. Turner, A study on real-time pricing electric tariffs. Energy Eng. **96**(5), 7–34 (1999)
3. W. Ketter, J. Collins, The 2013 power trading agent competition (2013)
4. M. Räsänen, J. Ruusunen, R.P. Hämäläinen, Optimal tariff design under consumer self-selection. Energy Econ. **19**(2), 151–167 (1997)
5. NIST, Smart grid: a begginer's guide (2012)
6. M. Maenhoudt, G. Deconinck, Agent-based modelling as a tool for testing electric power market designs, in *2010 7th International Conference on the European Energy Market* (2010), pp. 1–5
7. S. Borenstein, To what electricity price do consumers respond? residential demand elasticity under increasing-block pricing, in *Preliminary Draft April*, vol. 30 (2009)
8. P.P. Reddy, M. Veloso, Learned behaviors of multiple autonomous agents in smart grid markets (2011), pp. 1396–1401
9. C.J. Watkins, P. Dayan, Q-learning. Mach. Learn. **8**(3–4), 279–292 (1992)

Part II
Automated Negotiating Agents Competition

The Sixth Automated Negotiating Agents Competition (ANAC 2015)

Katsuhide Fujita, Reyhan Aydoğan, Tim Baarslag, Koen Hindriks, Takayuki Ito and Catholijn Jonker

Abstract In May 2015, we organized the Sixth International Automated Negotiating Agents Competition (ANAC 2015) in conjunction with AAMAS 2015. ANAC is an international competition that challenges researchers to develop a successful automated negotiator for scenarios where there is incomplete information about the opponent. One of the goals of this competition is to help steer the research in the area of multi-issue negotiations, and to encourage the design of generic negotiating agents that are able to operate in a variety of scenarios. 24 teams from 9 different institutes competed in ANAC 2015. This chapter describes the participating agents and the setup of the tournament, including the different negotiation scenarios that were used in the competition. We report on the results of the qualifying and final round of the tournament.

K. Fujita (✉)
Faculty of Engineering, Tokyo University of Agriculture and Technology, Tokyo, Japan
e-mail: katfuji@cc.tuat.ac.jp

R. Aydoğan
Computer Science Department, Özyeğin University, Istanbul, Turkey
e-mail: reyhan.aydogan@gmail.com

T. Baarslag
Agents, Interaction and Complexity group, University of Southampton, Southampton, UK
e-mail: tb1m13@ecs.soton.ac.uk

T. Ito
Techno-Business Administration (MTBA), Nagoya Institute of Technology, Aichi, Japan
e-mail: ito.takayuki@nitech.ac.jp

K. Hindriks · C. Jonker
Man Machine Interaction Group, Delft University of Technology, Delft, The Netherlands
e-mail: K.V.Hindriks@tudelft.nl

C. Jonker
e-mail: C.M.Jonker@tudelft.nl

© Springer International Publishing AG 2017
K. Fujita et al. (eds.), *Modern Approaches to Agent-based Complex Automated Negotiation*, Studies in Computational Intelligence 674,
DOI 10.1007/978-3-319-51563-2_9

1 Introduction

Success in developing an automated agent with negotiation capabilities has great advantages and implications. In order to help focus research on proficiently negotiating automated agents, we have organized the automated negotiating agents competition (ANAC). The results of the different implementations are difficult to compare, as various setups are used for experiments in ad hoc negotiation environments [6]. An additional goal of ANAC is to build a community in which work on negotiating agents can be compared by standardized negotiation benchmarks to evaluate the performance of both new and existing agents. Recently, the analysis of ANAC becomes important fields of automated negotiations in multi-agent systems [1].

In designing proficient negotiating agents, standard game-theoretic approaches cannot be directly applied. Game theory models assume complete information settings and perfect rationality [8, 9]. However, human behavior is diverse and cannot be captured by a monolithic model. Humans tend to make mistakes, and they are affected by cognitive, social and cultural factors [7]. A means of overcoming these limitations is to use heuristic approaches to design negotiating agents. When negotiating agents are designed using a heuristic method, we need an extensive evaluation, typically through simulations and empirical analysis.

We employ an environment that allows us to evaluate agents in a negotiation competition: GENIUS [6], a **G**eneral **E**nvironment for **N**egotiation with **I**ntelligent multi-purpose **U**sage **S**imulation. GENIUS helps facilitating the *design* and *evaluation* of automated negotiators' strategies. It allows easy development and integration of existing negotiating agents, and can be used to simulate individual negotiation sessions, as well as tournaments between negotiating agents in various negotiation scenarios. The design of general automated agents that can negotiate proficiently is a challenging task, as the designer must consider different possible environments and constraints. GENIUS can assist in this task, by allowing the specification of different negotiation domains and preference profiles by means of a graphical user interface. It can be used to train human negotiators by means of negotiations against automated agents or other people. Furthermore, it can be used to teach the design of generic automated negotiating agents.

The First Automated Negotiating Agents Competition (ANAC 2010) was held in May 2010, with the finals being run during the AAMAS 2010 conference. Seven teams had participated and three domains were used. *AgentK* generated by the Nagoya Institute of Technology team won the ANAC 2010 [2]. The Second Automated Negotiating Agents Competition (ANAC 2011) was held in May 2011, with the AAMAS 2011 conference. 18 teams had participated and eight domains were used. The new feature of ANAC 2011 was the discount factor. *HardHeaded* generated by the Delft University of Technology won the ANAC 2011 [3]. The Third Automated Negotiating Agents Competition (ANAC 2012) was held in May 2012, with the AAMAS 2012 conference. 17 teams had participated and 24 domains were used. The new feature of ANAC 2012 was the reservation value. *CUHKAgent* generated by the Chinese University of Hong Kong won the ANAC 2012 [10]. The Forth

Automated Negotiating Agents Competition (ANAC 2013) was held in May 2013, with the AAMAS 2013 conference. 19 teams had participated and 24 domains were used. The new feature of ANAC 2013 was that agents can use the bidding history. *The Fawkes* generated by the Delft University of Technology won the ANAC 2013 [5]. The Fifth Automated Negotiating Agents Competition (ANAC 2014) was held in May 2014, with the AAMAS 2014 conference. 21 teams had participated and 12 domains were used. The new feature of ANAC 2014 was nonlinear utility functions. *AgentM* generated by Nagoya Institute of Technology won the ANAC 2014 [4].

ANAC organizers have been employing some of the new feature every year to develop the ANAC competition and the automated negotiations communities. The challenge of ANAC 2015 is to reach an agreement while negotiating with two opponents at the same time. In addition,the utility functions are linear again, as they were in ANAC 2010–2013. The multi-player protocol is a simple extension of the bilateral alternating offers protocol, called the Stacked Alternating Offers Protocol for Multi-Lateral Negotiation (SAOP).

The timeline of ANAC 2015 is mainly consisted by two parts: Qualifying Round and Final Round. First, the qualifying round was played in order to select the finalists from 24 agents by considering the individual utility and the nash product. In the qualifying round, 24 agents was divided into four groups (pools) randomly, and the best two agents of those pools proceed to the final in each category. After that, the final round was played among 8 agents in two categories, which won the qualifying round. The domains and preference profiles in the qualifying and final rounds were 10 domains generated by the organizers. The entire matches played among 8 agents in each category, and the ranking of ANAC 2015 is decided.

The remainder of this chapter is organized as follows. Section 2 provides an overview over the design choices for ANAC, including the model of negotiation, tournament platform and evaluation criteria. In Sect. 3, we present the setup of ANAC 2015 followed by Sect. 4 that layouts the results of competition. Finally, Sect. 5 outlines our conclusions and our plans for future competitions.

2 Setup of ANAC 2015

2.1 Negotiation Model

Given the goals outlined in the introduction, in this section we introduce the set-up and negotiation protocol used in ANAC. The interaction between negotiating parties is regulated by a *negotiation protocol* that defines the rules of how and when proposals can be exchanged. The parties negotiate over a set of *issues*, and every issue has an associated range of alternatives or *values*. A negotiation *outcome* consists of a mapping of every issue to a value, and the set, Ω of all possible outcomes is called the negotiation *domain*. The domain is common knowledge to the negotiating parties and stays fixed during a single negotiation session. In addition to the domain, both parties also have privately-known preferences described by their *preference profiles* over Ω. These preferences are modeled using a utility function U that maps a possible outcomes $\omega \in \Omega$ to a real-valued number in the range [0, 1]. In ANAC 2015, the utilities are *linearly additive*. That is, the overall utility consists of a weighted sum of the utility for each individual issue. While the domain (i.e. the set of outcomes) is common knowledge, the preference profile of each player is private information. This means that each player has only access to its own utility function, and does not know the preferences of its opponent.[1] Moreover, we use the term *scenario* to refer to the domain and the pair of preference profiles (for each agent) combined.

Finally, we supplement it with a deadline, reservation value and discount factors. The reasons for doing so are both pragmatic and to make the competition more interesting from a theoretical perspective. In addition, as opposed to having a fixed number of rounds, both the discount factor are measured in *real time*. In particular, it introduces yet another factor of uncertainty since it is now unclear how many negotiation rounds there will be, and how much time an opponent requires to compute a counter offer. In ANAC 2015, the discount factors and reservation value depend on the scenario, but the deadline is set to three minutes. The implementation of discount factors in ANAC 2015 is as follows:

A negotiation lasts a predefined time in seconds*(deadline)*. The timeline is normalized, i.e.: time $t \in [0, 1]$, where $t = 0$ represents the start of the negotiation and $t = 1$ represents the deadline. When agents can make agreements in the deadline, the individual utilities of each agent are the *reservation value*. Apart from a deadline, a scenario may also feature *discount factors*. Discount factors decrease the utility of the bids under negotiation as time passes. Let d in [0, 1] be the discount factor. Let t in [0, 1] be the current normalized time, as defined by the timeline. We compute the discounted utility U_D^t of an outcome ω from the undiscounted utility function U as follows:

$$U_D^t(\omega) = U(\omega) \cdot d^t \tag{1}$$

At $t = 1$, the original utility is multiplied by the discount factor. Furthermore, if $d = 1$, the utility is not affected by time, and such a scenario is considered to be undiscounted.

In the competition, we use the Stacked Alternating Offers Protocol for Multi-Lateral Negotiation (SAOP) as the new feature, in which the negotiating parties exchange offers in turns. All of the participants around the table get a turn per round; turns are taken clock-wise around the table. The first party starts the negotiation with an offer that is observed by all others immediately. Whenever an offer is made the next party in line can take the following actions:

[1] We note that, in the competition each agent plays *all* preference profiles, and therefore it would be possible in theory to learn the opponent's preferences. However, the rules explicitly disallow learning *between* negotiation sessions, and only *within* a negotiation session. This is done so that agents need to be designed to deal with unknown opponents.

- Make a counter offer (thus rejecting and overriding the previous offer)
- Accept the offer
- Walk away (e.g. ending the negotiation without any agreement)

This process is repeated in a turn taking clock-wise fashion until reaching an agreement or reaching the deadline. To reach an agreement, all parties should accept the offer. If at the deadline no agreement has been reached, the negotiation fails. The details of SAOP is written in the next chapter.

2.2 Running the Tournament

As a tournament platform to run and analyze the negotiations, we use the GENIUS environment (General Environment for Negotiation with Intelligent multi-purpose Usage Simulation) [6]. GENIUS is a research tool for automated multi-issue negotiation, that facilitates the design and evaluation of automated negotiators' strategies. It also provides an easily accessible framework to develop negotiating agents via a public API. This setup makes it straightforward to implement an agent and to focus on the development of strategies that work in a general environment.

GENIUS incorporates several mechanisms that aim to support the design of a general automated negotiator. The first mechanism is an analytical toolbox, which provides a variety of tools to analyse the performance of agents, the outcome of the negotiation and its dynamics. The second mechanism is a repository of domains and utility functions. Lastly, it also comprises repositories of automated negotiators. In addition, GENIUS enables the evaluation of different strategies used by automated agents that were designed using the tool. This is an important contribution as it allows researchers to empirically and *objectively* compare their agents with others in different domains and settings.

The timeline of ANAC 2015 consists of two phases: the qualifying round and the final round. The domains and preference profiles used during the competition are not known in advance and were designed by the organizers. An agent's success is measured using the evaluation metric in all negotiations of the tournament for which it is scheduled.

First, a qualifying round was played in order to select the finalists from the 24 agents that were submitted by the participating teams. Since there were too many agents, in the different domains, a whole tournament in the qualifying round is impossible. Therefore, 24 agents was divided to four groups (pools) randomly, and the best two agents in nash product and individual utility in each pool proceed to the final round. It took two weeks to finish the all pools of the qualifying round. In ANAC-2015, we didn't allow the updating agents between the qualifying round and the final round.

The final round was played among the the agents that achieved the best scores (individual utility and nash product) in each pool during qualifying. We prepared two categories in the final round of ANAC 2015: Individual utility category and nash product categories. The domains and preference profiles are same as the qualifying round. The entire matches played among agents, and the final ranking of ANAC 2015 was decided. To reduce the effect of variation in the results, the final score calculates the average of the five trials.

3 Competition Domains and Agents

3.1 Scenario Descriptions

The ANAC is aimed towards modeling multi-issue negotiations in uncertain, open environments, in which agents do not know what the preference profile of the opponent is. The various characteristics of a negotiation scenario such as size, number of issues, opposition, discount factor and reservation value can have a great influence on the negotiation outcome. Therefore, we generated ten types of domains and profiles in the competition. Especially, in the qualifying round and final round, we used all 10 scenarios with different discount factors and reservation values and profiles. In other words, they have vary in terms of the number of issues, the number of possible proposals, the opposition of the preference profiles (see Table 1). The 3d negotiation space plotting in each domain are represented graphically in Fig. 1.

Table 1 The domains used in ANAC 2015

ID	Number of issues	Size	Discount factor	Reservation value	Cooperativeness
1	1	5	None	0.5	Very competitive
2	1	5	None	0.5	A bit collaborative
3	2	25	0.2	None	Very competitive
4	2	25	None	0.5	Quite collaborative
5	4	320	0.5	None	Competitive
6	4	320	0.5	None	Collaborative
7	8	3^8	None	None	Competitive
8	8	3^8	None	None	Collaborative
9	16	2^{16}	0.4	0.7	Very collaborative
10	16	2^{16}	0.4	0.7	Very competitive

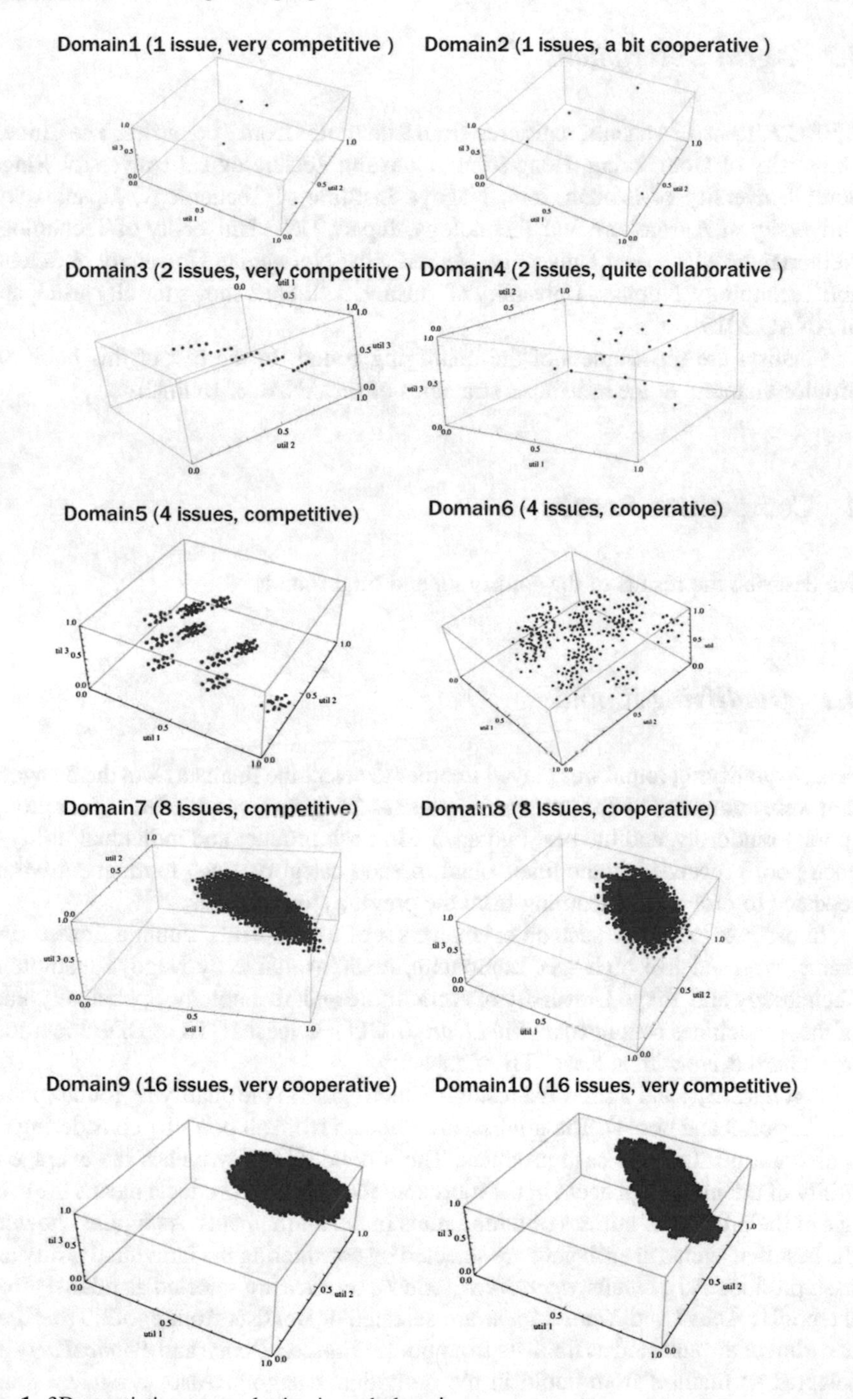

Fig. 1 3D negotiation space plotting in each domain

3.2 Agent Descriptions

ANAC 2015 had 24 agents, registered from 9 institutes from 7 countries: The Chinese University of Hong Kong, Hong Kong; Nanyang Technological University, Singapore; University of Isfahan, Iran; Nagoya Institute of Technology, Japan; Tokyo University of Agriculture and Technology, Japan; Delft University of Technology, Netherlands; Maastricht University, Netherlands; Norwegian University of Science and Technology, Norway; University of Tulsa, US. Table 2 shows the all participants in ANAC 2015.

Finalists are the winners of the qualifying round. In the rest of this book, we provide chapters of the individual strategies of the ANAC2015 finalists.

4 Competition Results

We describe the results of the qualifying and final rounds.

4.1 Qualifying Round

First, a qualifying round was played in order to select the finalists from the 24 agents that were submitted by the participating teams. 24 agents was divided to four groups (pools) randomly, and the best two agents in nash product and individual utility in each pool proceeded to the final round in each category. Each tournament wasn't repeated to prohibit the learning from the previous tournaments.

In order to complete such an extensive set of tournaments within a limited time frame, we used five high-spec computers, made available by Nagoya Institute of Technology and Tokyo University of Agriculture and Technology. Specifically, each of these machines contained an *Intel Core i7* CPU, at least 16GB of DDR3 memory, and a hard drive with at least 2TB of capacity.

Figures 2, 3, 4 and 5 show the results of each agent in the qualifying round (pool1, pool2, pool3 and pool4). The finalists are selected from all pools by considering the individual utilities and nash products. The individual utility means the average of utility of the individual agent in the tournaments. The nash products means the average of the product of utilities of three agents in the tournaments. As figures showing, the best two agents in each pool are selected by considering the individual utility and nash product. As a results, *agentBuyog* and *PokerFace* are selected as finalists from the pool1; *Atlas3* and *XianFaAgent* are selected as finalists from pool2; *ParsAgent* and *kawaii* are selected as finalists from pool3; *RandomDance* and *PhonexParty* are selected as finalists from pool4 in the individual category. Also, *agentBuyog* and *Mercury* are selected as finalists from the pool1; *Atlas3* and *AgentX* are selected as finalists from pool2; *CUHKAgent* and *Jonny Black* are selected as finalists from

Table 2 Team members and agent names in ANAC 2015

No.	Team members	Affliction	Agent name
1	Jeroen Peperkamp Vikko Smit	Delft University of Technology	Pokerface
2	Joao Almeida Hugo Zwaan Xiaoran Liu	Delft University of Technology	TUDMixedStrategyAgent
3	Dirk Schut Nikol Guljelmovic Jelle Munk	Delft University of Technology	Ai Caramba!
4	Shuang Zhou	Maastricht University	Mercury
5	Siqi Chen Jianye Hao Gerhard Weiss HF-Leung	Maastricht University	AresParty
6	Shinji Kakimoto	Tokyo University of Agriculture and Technology	RandomDance
7	Hiroyuki Shinohara	Tokyo University of Agriculture and Technology	AgentHP
8	Bhargav Sosale Swarup Satish Suyog Shivakumar Bo An	Nanyang Technological University	Agent Buyog
9	Neo Jun	Nanyang Technological University	AgentNeo
10	Chen Xian Fa Kelvin	Nanyang Technological University	XianFaAgent
11	Sengoku Akihisa	Nagoya Institute of Technology	SENGOKU
12	Ishida Kenta	Nagoya Institute of Technology	AgentW
13	Masayuki Hayashi	Nagoya Institute of Technology	Agent H - Hayashi
14	Bun Koku	Nagoya Institute of Technology	AgentX
15	Akiyuki Mori	Nagoya Institute of Technology	Atlas3
16	Takuma Inamoto	Nagoya Institute of Technology	Kawaii
17	Kazumasa Takahashi	Nagoya Institute of Technology	DragKnight
18	Zenefa Rahaman Kendall Hyatt Chad Crawford Sandip Sen	University of Tulsa	PNegotiator

(continued)

Table 2 (continued)

No.	Team members	Affliction	Agent name
19	Nathaniel Beckemeyer Samuel Beckmann Abigail Sislo	University of Tulsa	MeanBot
20	Osman Yucel Jon Hoffman	University of Tulsa	Jonny Black
21	Lam Wing	The Chinese University of Hong Kong	PhoenixParty
22	Leung Hoi Tang Ng Chi Wing Ho-fung Leung	The Chinese University of Hong Kong	CUHKAgent2015
23	Zahra Khosravimehr Faria Nasiri Mofakham	University of Isfahan	ParsAgent
24	Lars Liahagen Haakon H. Rod	Norwegian University of Science and Technology (NTNU)	Forseti

Agent Name	Average (Individual)	Rank (Individual)	Average (Nash Product)	Rank (Nash)
agentBuyogV2	**0.597955067**	**1**	**0.275919234**	**2**
PokerFace	**0.594266467**	**2**	0.268609115	5
PNegotiator	0.591739600	3	0.268958032	4
DrageKnight	0.571351533	4	0.272096395	3
Mercury	0.550937867	5	**0.300345722**	**1**
SENGOKU	0.547276433	6	0.263140705	6

Fig. 2 Average scores of each agent in the qualifying round (pool1)

Agent Name	Average (Individual)	Rank (Individual)	Average (Nash Product)	Rank (Nash)
Atlas3	**0.680664517**	**1**	**0.309788773**	**1**
XianFaAgent	**0.633863800**	**2**	0.288605994	3
MeanBot	0.584072250	3	0.223485414	6
AgentX	0.571492517	4	**0.290723081**	**2**
AgentHP	0.535089883	5	0.254176388	5
TUDMixed	0.504430117	6	0.284713164	4

Fig. 3 Average scores of each agent in the qualifying round (pool2)

pool3; *RandomDance* and *AgentH* are selected as finalists from pool4 in the nash product category.

Agent Name	Average (Individual)	Rank (Individual)	Average (Nash Product)	Rank (Nash)
ParsAgent	**0.582228250**	1	0.226744176	6
kawaii	**0.575404450**	2	0.224477192	5
Group2	0.567122400	3	0.230310631	4
CUHKAgent2015	0.552638067	4	**0.268464466**	1
AgentW	0.518159433	5	0.237260891	3
JonnyBlack	0.491797117	6	**0.261849645**	2

Fig. 4 Average scores of each agent in the qualifying round (pool3)

Agent Name	Average (Individual)	Rank (Individual)	Average (Nash Product)	Rank (Nash)
RandomDance	**0.408558450**	1	**0.182175298**	1
PhoenixParty	**0.380885900**	2	0.136591820	6
AresParty	0.378801767	3	0.147792872	4
AgentNeo	0.356815667	4	0.141631385	5
AgentH	0.339627333	5	**0.163050174**	2
Forseti	0.258990217	6	0.151909061	3

Fig. 5 Average scores of each agent in the qualifying round (pool4)

4.2 *Final Round*

It is notable that *Atlas3* was the clear winner of the both categories (see Tables 3 and 4). However, the differences in utilities between many of the ranked strategies are small, so several of the agents were decided the ranking by a small margin. Finally, the first places in the individual utility and nash product categories were awarded to *Atlas3* ($450); The second place in the individual category was awarded to the ParsAgent ($150); The second place in the nash product was awarded awarded to *Mercury* ($175); The third place in the individual category was awarded to *RandomDance*

Table 3 Tournament results in the final round (Individual utility)

Rank	Agent	Score	Standard deviation
1	Atlas3	0.481042722	0.00156024
2	ParsAgent	0.470693979	0.003128712
3	RandomDance	0.46062548	0.003038258
4	kawaii	0.460129481	0.002715924
5	agentBuyog	0.458823101	0.003842303
6	PhoenixParty	0.442975836	0.005032221
7	XianFaAgent	0.353133027	0.001918821
8	PokerFace	0.344003999	0.001433044

Table 4 Tournament results in the final round (Nash product)

Rank	Agent	Score	Standard deviation
1	Atlas3	0.323992201	0.000405256
2	Mercury	0.321600864	0.001620108
3	JonnyBlack	0.313749314	0.001026152
4	AgentX	0.312427823	0.001393852
5	CUHKAgent	0.309464847	0.001726555
6	RandomDance	0.294950885	0.001088483
7	AgentH	0.292136808	0.001547118
8	agentBuyog	0.282378625	0.00236416

($100); The third place in the nash product was awarded awarded to *JonnyBlack* ($125).

5 Conclusion

This chapter describes the Sixth automated negotiating agents competition (ANAC2015). Based on the process, the submissions and the closing session of the competition we believe that our aim has been accomplished. Recall that we set out for this competition in order to steer the research in the area multi-issue closed negotiation. 24 teams have participated in the competition and we hope that many more will participate in the following competitions.

ANAC also has an impact on the development of GENIUS. We have released a new, public build of GENIUS[2] containing all relevant aspects of ANAC. In particular, this includes all domains, preference profiles and agents that were used in the competition. This will make the complete setup of ANAC available to the negotiation research community. Not only have we learnt from the strategy concepts introduced in ANAC, we have also gained understanding in the correct setup of a negotiation competition. The joint discussion with the teams gives great insights into the organizing side of the competition.

To summarize, the agents developed for ANAC will proceed the next step towards creating autonomous bargaining agents for real negotiation problems. We plan to organize the next ANAC in conjunction with the next AAMAS conference.

Acknowledgements The authors would like to thank the team of masters students at Nagoya Institute of Technology, Japan for their valuable help in the organization of the ANAC 2015 competition.

[2] http://ii.tudelft.nl/genius.

References

1. T. Baarslag, K. Fujita, E.H. Gerding, K.V. Hindriks, T. Ito, N.R. Jennings, C.M. Jonker, S. Kraus, R. Lin, V. Robu, C.R. Williams, Evaluating practical negotiating agents: results and analysis of the 2011 international competition. Artif. Intell. **198**, 73–103 (2013)
2. T. Baarslag, K.V. Hindriks, C.M. Jonker, S. Kraus, R. Lin, The first automated negotiating agents competition (ANAC 2010), in *New Trends in Agent-Based Complex Automated Negotiations* (Springer, Heidelberg, 2012), pp. 113–135
3. K. Fujita, T. Ito, T. Baarslag, K.V. Hindriks, C.M. Jonker, S. Kraus, R. Lin, The second automated negotiating agents competition (ANAC2011), in *Complex Automated Negotiations: Theories, Models, and Software Competitions* (Springer, Heidelberg, 2013), pp. 183–197
4. N. Fukuta, T. Ito, M. Zhang, K. Fujita, V. Robu, The fifth automated negotiation competition, in *Recent Advances in Agent-Based Complex Automated Negotiation* (Springer, Switzerland, 2016)
5. Y.K. Gal, L. Ilany, The fourth automated negotiation competition, in *Next Frontier in Agent-Based Complex Automated Negotiation* (Springer, Japan, 2015), pp. 129–136
6. R. Lin, S. Kraus, T. Baarslag, D. Tykhonov, K.V. Hindriks, C.M. Jonker, Genius: an integrated environment for supporting the design of generic automated negotiators. Comput. Intell. **30**(1), 48–70 (2014)
7. R.D. McKelvey, T.R. Palfrey, An experimental study of the centipede game. Econometrica **60**(4), 803–36 (1992)
8. M. Osborne, A. Rubinstein, in *Bargaining and Markets*, Economic Theory, Econometrics, and Mathematical Economics (Academic Press, Cambridge, 1990)
9. M.J. Osborne, A. Rubinstein, in *A Course in Game Theory*, MIT Press Books (MIT Press, Cambridge, 1994)
10. C.R. Williams, V. Robu, E.H. Gerding, N.R. Jennings, An overview of the results and insights from the third automated negotiating agents competition (ANAC2012), in *Novel Insights in Agent-based Complex Automated Negotiation* (Springer, Japan, 2014), pp. 151–162

Alternating Offers Protocols for Multilateral Negotiation

Reyhan Aydoğan, David Festen, Koen V. Hindriks and Catholijn M. Jonker

Abstract This paper presents a general framework for multilateral turn-taking protocols and two fully specified protocols namely Stacked Alternating Offers Protocol (SAOP) and Alternating Multiple Offers Protocol (AMOP). In SAOP, agents can make a bid, accept the most recent bid or walk way (i.e., end the negotiation without an agreement) when it is their turn. AMOP has two different phases: *bidding* and *voting*. The agents make their bid in the bidding phase and vote the underlying bids in the voting phase. Unlike SAOP, AMOP does not support walking away option. In both protocols, negotiation ends when the negotiating agents reach a joint agreement or some deadline criterion applies. The protocols have been evaluated empirically, showing that SAOP outperforms AMOP with the same type of conceder agents in a time-based deadline setting. SAOP was used in the ANAC 2015 competition for automated negotiating agents.

Keywords Multilateral negotiation · Turn-taking negotiation protocol · Alternating offers protocol

R. Aydoğan (✉) · D. Festen · K.V. Hindriks · C.M. Jonker
Interactive Intelligence Group, Delft University of Technology, Delft, The Netherlands
e-mail: R.Aydogan@tudelft.nl; reyhan.aydogan@ozyegin.edu.tr

D. Festen
e-mail: D.Festen@tudelft.nl

K.V. Hindriks
e-mail: K.V.Hindriks@tudelft.nl

C.M. Jonker
e-mail: C.M.Jonker@tudelft.nl

R. Aydoğan
Computer Science Department, Özyeğin University, Istanbul, Turkey

© Springer International Publishing AG 2017
K. Fujita et al. (eds.), *Modern Approaches to Agent-based Complex Automated Negotiation*, Studies in Computational Intelligence 674,
DOI 10.1007/978-3-319-51563-2_10

1 Introduction

Multilateral negotiation is an important form of group decision making [2, 6]. In many aspects of life, whether in a personal or a professional context, consensus decisions have to be made (e.g., setting the agenda in a business meeting, and the time, and location of the meeting). The complexity of multilateral negotiation increases with the number of negotiating parties [16], and with the complexity of the negotiation domain (see, e.g., [13]). The more complex the negotiations, the more human negotiators may have difficulty in finding joint agreements and the more they might benefit from the computational power of automated negotiation agents and/or negotiation support tools.

For bilateral negotiation the main challenges are opponent modeling, bidding and acceptance strategies have been extensively studied in the multi-agent community [7]. The brunt of the work is based on the alternating offers protocol to govern the interaction between negotiating agents. According to this protocol, one of the negotiating parties starts the negotiation with an offer. The other party can either accept or reject the given offer. By accepting the offer, the negotiation ends with an agreement. When rejecting the offer the other party can either end the negotiation (walk away), or make a counter offer. This process continues in a turn-taking fashion.

This paper presents a general framework for multilateral turn-taking negotiation protocols, in which fundamental definitions and rules are described formally. Based on this formal framework, we define two negotiation protocols, namely *Stacked Alternating Offers Protocol* (SAOP) and *Alternating Multiple Offers Protocol* (AMOP). In both protocols, negotiating agents can only take their action when it is their turn, the turn taking sequences are defined before the negotiation starts. SAOP allows negotiating agents to evaluate only the most recent bid in their turn and accordingly they can either accept that bid or make a counter offer or walk away. By contrast, in AMOP all agents bid sequentially and then, they vote on all bids iteratively (i.e., either accept or reject). Consequently, agents can see each agent's opinion on their bid. As a result, in AMOP the agents have a better overview of the outcome space (e.g., which bids are acceptable or not acceptable for their opponents). On the other hand, the communication cost is higher in contrast to the stacked alternating offers protocol. SAOP was used in the ANAC 2015 competition for automated negotiating agents that was organized to facilitate the research on multilateral negotiation. AMOP was developed as an alternative in which agents can get more information from their opponents by getting votes from all agents on all bids made.

To see how well the agents perform in each protocol and to judge the fairness of the outcomes, we implemented both protocols in Genius and compared them empirically. The current results show that SAOP outperforms AMOC on the given negotiation scenarios with respect to the social welfare criterion.

The rest of this paper is organized as follows. Section 2 presents the general framework for multilateral turn-taking protocols. The stacked alternating offers protocol and alternating multiple offers protocol are explained in Sects. 3 and 4 respectively. Section 5 explains our experimental setup, metrics and results. Section 6 discusses the related work. Finally, we conclude the paper with directions to future work in Sect. 7.

2 Formal Framework for Multilateral Turn-Taking Protocols

Before presenting two variants of turn-taking protocols for multilateral negotiation, we first introduce a general formal framework for specifying these protocols. The framework consists of a number of general definitions regarding alternating offers protocols for multilateral negotiations. In later sections where we present two turn-taking protocols for multilateral negotiation, those concepts that are protocol dependent will be revisited.

2.1 Basic Notation

The basic notions of a negotiation are the agents that negotiate, the bids that they exchange, and the other actions that they can take during the negotiation. We use Agt to denote a finite set of agent names, Bid to denote a set of bids over the negotiation domain, and $Act \subseteq Bid \cup \{accept, reject, end\}$ to denote a set of possible actions that can be taken during the negotiation where end denotes that the agent walks away. In this document tuples and sequences are used frequently. For any tuple or sequence t and any index $i \in \mathbb{N}$, let t_i denote the ith element of tuple t, and similarly, for any tuple, sequence or set t, let $|t|$ denote the number of elements in t.

Definition 1 Round and Phase.
Rounds and phases within rounds are used to structure the negotiation process. Although the structure of the phases differs over protocols, the concepts are defined generally as follows:

- $Round \subseteq \mathbb{N}^+$ is the set of round numbers. Rounds are numbered from 1 onwards, if i is the current round, then the next round is numbered $i + 1$.
- $Phase \subseteq \mathbb{N}$ is the set of phase identifiers. Phases are numbered from 0 onwards, if i is the current phase, then the next phase is numbered $i + 1$. The set $Phase$ can be a singleton. Let ℓ denote the last phase, which is equal to $|Phase| - 1$.
- $RPhase = Round \times Phase$, the first argument denotes the round number whereas the second argument denotes the specific phase of that round. This depends on the protocol at hand. In case $Phase$ is $\{ 0 \}$, then, for convenience, $RPhase$ is collapsed to $Round$ only.

Definition 2 Turn taking.
Alternating offer protocols assign turns to the negotiating agents. Turns are taken according to a turn-taking sequence.

- $TurnSeq = Agt^{|Agt|}$ is a sequence of agents, such that
  - $\forall\, s \in TurnSeq\ \forall\, a \in Agt, \exists i \in \mathbb{N}^+, i \leq |s|$ such that $s_i = a$ and
  - $\forall\, s \in TurnSeq\ \forall\, i, j \leq |s|$: $s_i{=}s_j \rightarrow i{=}j$.

- The function $rpSeq : RPhase \rightarrow TurnSeq$ assigns a turn-taking sequence per round and phase. Its specification depends on the protocol.
- The function $prev : RPhase \times \mathbb{N}^+ \rightarrow RPhase \times \mathbb{N}^+$ defines the previous turn in the negotiation, that can be in this round-phase or a previous round-phase, specified by:

$$prev(r,t) = \begin{cases} \langle r, t-1 \rangle, & 1 < t \leq |Agt| \\ \langle \langle r_1, r_2 - 1 \rangle, |Agt| \rangle, & t = 1 \wedge r_2 > 0 \\ \langle \langle r_1 - 1, \ell \rangle, |Agt| \rangle, & t = 1 \wedge r_2 = 0 \wedge r_1 > 1 \\ \text{undefined}, & \text{otherwise} \end{cases} \quad (1)$$

To be able to specify what happened $k \in \mathbb{N}$ turns ago, we recursively define $prev^k : RPhase \times \mathbb{N}^+ \rightarrow RPhase \times \mathbb{N}^+$ as follows:

$$\forall x \in RPhase \times \mathbb{N}^+ :$$
$$prev^0(x) = x$$
$$prev^1(x) = prev(x)$$
$$prev^{n+1}(x) = prev^n(prev(x))$$

The conditions ensure fairness in protocols in the sense that every agent gets a turn and no agent gets more than one turn in a sequence. In case the same turn-taking sequence is used in all rounds and phases, this sequence is denoted by s. This is true for the protocols SAOP and AMOP of the later sections. However, Definition 2 allows more freedom.

Although the actions might differ over protocols, we introduce notions that are general to all negotiation protocols.

Definition 3 Actions and allowed actions.
The functions *action* and *allowedAction* specify what actions agents take and what actions they are allowed to take.

- $action$: $Agt \times RPhase \rightarrow Act$. The term $action(a, r)$ denotes what action agent $a \in Agt$ took in round-phase $r \in RPhase$.
- *allowedAct*: $RPhase \times \mathbb{N}^+ \rightarrow \mathcal{P}(Act)$. The function determines the allowed actions per turn t at a given round-phase r. The function specification varies over protocols.

Although protocols do not specify what actions agents take during the negotiation, the function $action$ is defined here, as the type action taken by the agents do have an effect on the procedure as specified in Definitions 5, and 6.

Definition 4 Deadline.
$d : RPhase \times \mathbb{N}^+$ is a predicate that denotes whether or not the negotiation deadline has been reached. Its value is determined at the end of the current turn. Its specification depends on the protocol.

Examples of such criteria are round-based ($r > R_{deadline}$), and time-based ($time > T_{deadline}$).

Definition 5 Agent ending the negotiation.
The predicate $endP : RPhase \times \mathbb{N}^{+}$ denotes whether or not an agent has ended the negotiation. Its value is determined at the end of the current turn.

$$\forall r \in RPhase\ , \forall t \in \mathbb{N}^{+} : \text{endP}(r, t) \leftrightarrow \text{action}(rpSeq(r)_t, r) = end \tag{2}$$

Note that, in typical protocols, the negotiation terminates as soon as one of the negotiators walks away, i.e., takes the action *end*. However, there might be protocols in which the other negotiators might continue. In that case Definition 7 that determines whether a negotiation continues will have to be adapted.

Definition 6 Agreement.
For use in the next predicates and functions two predicates are introduced to identify when an agreement has been reached and what that agreement is.

- The predicate *agr*: $RPhase \times \mathbb{N}^{+}$ denotes whether or not an agreement is reached. Its value is determined at the end of the current turn. The exact specification varies over protocols.
- The predicate $agrB : Bid \times RPhase \times \mathbb{N}^{+}$ denotes the bid that was agreed on.

Definition 7 Continuation.
The predicate *cont*: $RPhase \times \mathbb{N}^{+}$ denotes whether the negotiation continues after the current turn. Its value is determined at the end of the current turn.

$$\forall r \in RPhase\ \forall t \in \mathbb{N}^{+} : cont(r, t) \leftrightarrow \neg d(r, t) \wedge \neg endP(r, t) \wedge \neg agr(r, t) \tag{3}$$

Definition 8 Outcome of the negotiation.
The function *outcome*: $Round \times \mathbb{N}^{+} \rightarrow Bid \cup \{fail\}$ that determines the negotiation outcome at the end of the current turn.

$$outcome(r, t) = \begin{cases} undefined, & cont(r, t) \\ fail, & \neg cont(r, t) \wedge \neg agr(r, t) \\ b, & t > 0 \wedge \neg cont(r, t) \wedge\ agrB(b, r, t) \end{cases} \tag{4}$$

Definition 9 Turn-taking Negotiation protocol.
A turn-taking negotiation protocol P is a tuple $\langle\ Agt,\ Act,\ Rules\ \rangle$ where Agt denotes the set of agents participating in the negotiation, Act is the set of possible actions the agents can take, and $Rules$ is the set of rules that specify the particulars of the protocol. It contains the following rules, or specializations thereof.

1. Turn-taking Rule 1: Each agent gets turns according to the turn taking sequences of the protocol as specified by the definitions for rounds, phases, and turn-taking.
2. Turn-taking Rule 2: There is no turn after the negotiation has terminated, according to the Termination Rule.

3. Actions Rule 1: The agents can only act in their turn, as specified by the Turn-taking Rules.
4. Actions Rule 2: The agents can only perform actions that are allowed at that moment, as specified by the definitions for allowed actions.
5. Termination Rule: The negotiation is terminated after round-phase r and turn t if $\neg cont(r, t)$, as defined by the definitions for continuation, agreement, deadline and agent ending the negotiation.
6. Outcome Rule: The outcome of a negotiation is determined by the definitions for outcome and agreement.

The above definitions form the core of a formal framework for multilateral turn-taking negotiation protocols. There are different ways to extend the bilateral alternating offers protocol to the multilateral case. The next sections introduce two variants of this protocol: Stacked Alternating Offers Protocol (Sect. 3) and Alternating Multiple Offers Protocol (Sect. 4). Both protocols are specified by providing the detailed descriptions of those predicates and functions that are protocol dependent.

3 Stacked Alternating Offers Protocol (SAOP)

According to this protocol, all of the participants around the table get a turn per round; turns are taken clock-wise around the table, also known as a Round Robin schedule [14]. One of the negotiating parties starts the negotiation with an offer that is observed by all others immediately. Whenever an offer is made, the next party in line can take the following actions:

- Accept the offer
- Make a counter offer (thus rejecting and overriding the previous offer)
- Walk away (thereby ending the negotiation without any agreement)

This process is repeated in a turn-taking clock-wise fashion until reaching a termination condition is met. The termination condition is met, if a unanimous agreement or a deadline is reached, or if one of the negotiating parties ends the negotiation. Formally, the *Stacked Alternating Offer Protocol* is defined by the following definitions. We only provide an instantiated version of those definitions that are protocol dependent, i.e., phases of the negotiation, turn taking, actions and allowed actions, agreement, and the rules of encounter. Note that we only specify what changed in those definitions with respect to Sect. 2. SAOP can work with any deadline, or no deadline at all.

Definition 10 Round and Phase (Definition 1 for SAOP).
The concept of Round is not changed, there is only one phase per round in SAOP, i.e., $Phase = \{0\}$.

Definition 11 Turn taking (Definition 2 for SAOP).
In SAOP the same turn taking sequence is used in all rounds. Let s denote that sequence, thus for SAOP the set of turn-taking sequences is $TurnSeq = \{s\}$.

The rules for turn taking are those specified in Definition 2, i.e., each agent gets exactly one turn per round, as specified by s. Note that, since there is only one phase per round, instead of mentioning phases per round, in SAOP only rounds are mentioned.

Definition 12 Actions and allowed actions (Definition 3 for SAOP).
The function *action* is unchanged. The detailed specification of $allowedAction$: $RPhase \times \mathbb{N}^+ \rightarrow Act$ is as follows:

$$allowedAct(r,t) = \begin{cases} Bid \cup \{\text{end}\}, & \text{if } cont(r,t) \wedge t = 1 \wedge r_1 = 1 \\ Bid \cup \{\text{accept, end}\}, & \text{if } cont(r,t) \wedge (t \neq 1 \vee r_1 \neq 1) \\ \emptyset, & \text{otherwise} \end{cases}$$

Definition 13 Deadline (Definition 4 for SAOP).
Predicate $d : RPhase \times \mathbb{N}^+$ denotes whether or not the negotiation deadline has been reached. Its value is determined at the end of the current turn according to the following.

$$\forall r \in RPhase\ \forall t \in \mathbb{N}^+ : \mathrm{d}(r,t) \leftrightarrow \text{currenttime} - \text{negostarttime} \geq \text{maxnegotime} \tag{5}$$

The variables *negostarttime* and *maxnegotime* are set per negotiation. For example in the ANAC 2015 competition, the variables *currenttime* and *negostarttime* were taken from the system time of the computer running the tournament, and *maxnegotime* was set at 3 min.

Definition 14 Agreement (Definition 6 for SAOP).
The predicate *agr*: *RPhase* $\times \mathbb{N}^+$ denotes whether or not an agreement is reached. The predicate *agrB : Bid* $\times$ *RPhase* $\times \mathbb{N}^+$ denotes the bid that was agreed on. Their values are determined at the end of turn. Their specifications are as follows.

$$\forall r \in RPhase, \forall t \in \mathbb{N}^+ : agr(r,t) \leftrightarrow$$
$$action(s_{prev_2^{|Agt|-1}(r,t)}, prev_1^{|Agt|-1}(r,t)) \in Bid \wedge$$
$$\forall 0 \leq i \leq |Agt| - 2 : action(s_{prev_2^i(r,t)}, prev_1^i(r,t)) = accept$$

$$\forall r \in RPhase, \forall t \in \mathbb{N}^+ :$$
$$agrB(action(s_{prev_2^{|Agt|-1}(r,t)}, prev_1^{|Agt|-1}(r,t)), r, t) \leftrightarrow cont(r,t) \wedge agr(r,t)$$

Informally, we have an agreement iff $|Agt| - 1$ turns previously, an agent made a bid that was subsequently accepted by all the other agents. The agent that made the bid, in the SAOP protocol, is assumed to find its own bid acceptable. In $agrB$ that bid that was made $|Agt| - 1$ turns ago, is set to be the agreed bid in the current round-phase and turn.

3.1 Example

Assume that there are three negotiating negotiation parties, a_1, a_2 and a_3. Agent a_1 starts the negotiation with an bid b_1. Agent a_2 can accept this bid, make a counter offer or walk way. Let assume that she decides to make a counter bid (b_2). Assume that agents a_3 and a_1 accept this offer. As they all agree on this bid (i.e. b_2 made by a_2 in the previous round), the negotiation ends, and the outcome is bid b_2.

4 Alternating Multiple Offers Protocol (AMOP)

The AMOP protocol is an alternating offers protocol in which the emphasis is that all players will get the same opportunities with respect to bidding. That is, all agents have a bid from all agents available to them, before they vote on these bids. This implemented in the following way: The AMOP protocol has a bidding phase followed by voting phases. In the bidding phase **all** negotiators put their offer on the table. In the voting phases all participants vote on all of the bids on the negotiation table. If one of the bids on the negotiation table is accepted by all of the parties, then the negotiation ends with this bid. This is an iterative process continuing until reaching an agreement or reaching the deadline. The essential difference with the SAOP protocol, is that the players do not override each others offers and the agents can take all offers into account before they vote on the proposals. From an information theoretical point of view, this is a major difference. The specification of this protocol asks for detailed specifications of the protocol dependent definitions, i.e., on round-phases, turn taking, actions and allowed actions, agreement, and the rules of encounter. Only the changes are specified.

Definition 15 Round and Phase (Definition 1 for AMOP).
The concept of Round is not changed. Protocol AMOP has one bidding phase, followed by $|Agt|$ voting phases, i.e., $Phase = \{0, 1, \ldots, |Agt|\}$ where 0 denotes the bidding phase while for each $i \in [1, |Agt|]$, i denotes the voting phase on the bid made in the ith turn.

Definition 16 Turn taking (Definition 2 for AMOP).
In AMOP the same turn taking sequence is used at each phase of all rounds. Let s denote that sequence, i.e., $TurnSeq = \{s\}$.

Definition 17 Actions and allowed actions (Definition 3 for AMOP).
We define the set of possible actions as Act=$Bid \cup \{accept, reject\}$. The function *action* is unchanged. The detailed specification of $allowedAction$: $RPhase \times \mathbb{N}^+ \rightarrow Act$ is as follows:

$$allowedAct(r, t) = \begin{cases} Bid, & \text{if } cont(r, t) \wedge r_2 = 0 \\ \{accept, reject\}, & \text{if } cont(r, t) \wedge r_2 > 0 \\ \emptyset, & \text{otherwise.} \end{cases}$$

All rounds starts with a bidding phase during which all agents make a bid in turn specified by the turn sequence. The bidding phase is followed by a voting phase for each bid on the table. This means that all agents first vote on the first bid that was put on the table in this round, then all votes for the second bid and so on. During each voting phase, agents take their turn according to turn taking sequence as defined by the turn taking rules. During the voting phases, agents can only accept or reject bids. That the votes in phase i, refer to the ith bid in the bidding phase is specified indirectly by Definition 18.

Definition 18 Agreement (Definition 6 for AMOP).
The predicate *agr*: $RPhase \times \mathbb{N}^+$ denotes whether or not an agreement is reached. The predicate $agrB : Bid \times RPhase \times \mathbb{N}^+$ denotes the bid that was agreed on. Their values are determined at the end of turn in voting phases. Their specifications are as follows.

$$\forall r \in RPhase\, \forall t \in \mathbb{N}^+ : agr(r, t) \leftrightarrow$$
$$r_2 > 0 \wedge t = |Agt| \wedge action(s_{r_2}, \langle r_1, 0\rangle) \in Bid \wedge \forall 1 \leq i \leq t : action(s_i, r) = accept$$

$$\forall r \in RPhase\, \forall t \in \mathbb{N}^+ : agrB(action(s_{r_2}, \langle r_1, 0\rangle), r, t) \leftrightarrow cont(r, t) \wedge agr(r, t)$$

In other words, we have an agreement at the ith phase of a given round-phase r, iff all agents in that round voted to accept the bid made by the ith agent in the turn taking sequence s.

Definition 19 Continuation (Definition 7 for AMOP).
The predicate *cont*: $RPhase \times \mathbb{N}^+$ denotes whether the negotiation continues after the current turn. Its value is determined at the end of the current turn.

$$\forall r \in RPhase\, \forall t \in \mathbb{N}^+ : cont(r, t) \leftrightarrow \neg d(r, t) \wedge \neg agr(r, t) \tag{6}$$

4.1 Illustration

In $Phase = 0$, all players put an offer on the table (b_1 by a_1, b_2 by a_2 etc.). Note that there is no restriction on the bids; agents are allowed to make the same bid as others, or the same bid they made before. In the $Phase = 1$, all agents vote for the bid made by a_1, in $Phase = 2$, they all vote for the bid made by a_2 and so on. When all agents accept a bid during a voting phase, negotiation ends with this bid. Suppose that all agents, for example, vote to accept bid b_2, then the negotiation terminates at the end of phase 2 of round 1. If there were more than 2 agents, then this implies that the agents don't vote anymore for bid b_3.

5 Experimental Evaluation

In order to compare the performance of SAOP and AMOP empirically, we incorporated these two protocols into the GENIUS [15] negotiation platform, that was developed to enable researchers to test and compare their agents in various settings. GENIUS serves as a platform for the annual Automated Negotiating Agents Competition (ANAC) [3]. Our extension enables GENIUS to run multilateral negotiations; subsequently, the challenge of the ANAC 2015 competition was chosen to be multilateral negotiation.

A state-of-the-art agent, *Conceder* agent has been adapted for both multilateral protocols. This agent calculates a target utility and makes an arbitrary bid within a margin of 0.05 of this target utility. The target utility is calculated as $targetUtil(t) = 1 - t^{0.5}$ where $0 \geq t \geq 1$, t is the remaining time. This formula is derived from the general form proposed in [8]. In this paper, we adopted the ANAC 2015 setup, where three negotiating agents negotiate to come to an agreement within a three-minute deadline. We generated 10 different negotiation scenarios for three parties. Agent preferences are represented by means of additive utility functions. The size of the negotiation domains ranges from 216 to 2304.

To investigate the impact of the degree of conflict on the performance of the negotiation protocols, the scenarios tested in our experiment are chosen in such a way that half of those scenarios are collaborative and the rest are competitive. In competitive scenarios, there are relatively less outcomes which make everyone happy. We ran each negotiation ten times. Each agent negotiates for each preference profile in different order; that results 600 negotiations in total per each protocol (6 ordering permutations of 3 agents $\times$ 10 scenarios $\times$ 10 times).

We evaluated the protocols in term of the fairness of their negotiation outcome and social welfare. For social welfare, we picked the well known utilitarian social welfare metric [6], which is the sum of the utilities gained by each agent at the end of a negotiation. For fairness, we adopt the product of the utilities gained by each agents [12]. Recall that the Nash solution is the negotiation outcome with the maximum product of the agent utilities. Table 1 shows the average sum and product of the agent utilities with their standard deviation over 60 negotiations per each negotiation scenario. It is worth noting that the first five negotiation scenarios are cooperative and the last five scenarios are competitive. As expected, the negotiations resulted in higher sum and product of utilities when the negotiation scenarios are cooperative.

When we compare the performance of two protocols with time-based conceder agents in terms of social welfare, it is obviously seen that on average SAOP outperformed AMOP in all scenarios. However, the average social welfare difference between two protocols is higher in cooperative negotiation scenarios compared to the competitive scenarios. We have similar results when we look at the average product of agent utilities. The distinction between cooperative and competitive scenarios became more visible for the product of agent utilities since there are a few outcomes that can make everyone happy. The agents gained higher product of utilities when

Table 1 Social welfare and Nash product for cooperative domains (Scenario 1–5) and competitive domains (Scenario 6–10). All intervals are 95% confidence intervals. Sample size: $N = 60$

	Social welfare			Nash product			Distance to nash	
	SAOP	AMOP	Δ	SAOP	AMOP	Δ	SAOP	AMOP
Scenario 1	2.74 ± 0.01	2.44 ± 0.06	0.30	0.76 ± 0.01	0.54 ± 0.04	0.22	0.00 ± 0.04	0.23 ± 0.31
Scenario 2	2.36 ± 0.00	2.01 ± 0.06	0.35	0.48 ± 0.00	0.30 ± 0.03	0.18	0.11 ± 0.05	0.33 ± 0.26
Scenario 3	2.60 ± 0.00	2.38 ± 0.05	0.22	0.65 ± 0.00	0.50 ± 0.03	0.15	0.00 ± 0.00	0.18 ± 0.29
Scenario 4	2.74 ± 0.00	2.53 ± 0.06	0.21	0.76 ± 0.00	0.60 ± 0.04	0.16	0.00 ± 0.01	0.17 ± 0.34
Scenario 5	2.89 ± 0.00	2.80 ± 0.03	0.09	0.90 ± 0.00	0.81 ± 0.02	0.09	0.07 ± 0.00	0.12 ± 0.15
Scenario 6	2.20 ± 0.01	1.90 ± 0.05	0.30	0.39 ± 0.01	0.25 ± 0.02	0.14	0.00 ± 0.22	0.27 ± 0.23
Scenario 7	1.73 ± 0.01	1.59 ± 0.04	0.14	0.19 ± 0.00	0.14 ± 0.01	0.05	0.25 ± 0.06	0.38 ± 0.29
Scenario 8	2.19 ± 0.00	2.11 ± 0.02	0.08	0.39 ± 0.00	0.35 ± 0.01	0.04	0.06 ± 0.03	0.17 ± 0.17
Scenario 9	2.03 ± 0.00	1.96 ± 0.03	0.07	0.31 ± 0.00	0.26 ± 0.02	0.05	0.14 ± 0.01	0.25 ± 0.33
Scenario 10	2.06 ± 0.01	2.00 ± 0.03	0.03	0.32 ± 0.00	0.29 ± 0.01	0.06	0.14 ± 0.03	0.26 ± 0.24

they followed SAOP. Similarly, the negotiation outcomes in SAOP are closer to the Nash solution compared to the outcomes in AMOP. Based on the statistical t-test on both the average sum and product of agent utilities, it can be concluded that the results for SAOP with *Conceder* agent are statistically significantly better than the results for AMOP with *Conceder* agent on the given negotiation scenarios ($p \ll 0.001$).

The potential reasons why the social welfare of the agents are higher in SAOP compared to AMOP although they employ the same *Conceder* strategy stem from the main differences between SAOP and AMOP. One of these is that according to SAOP, the agents evaluate only the most recent bid in their turn whereas in AMOP, they evaluate all bids made by all agents in the current round. Although it sounds more fair to evaluate all bids made by all, the agents do not obtain a more fair outcome in AMOP. This may stem from the fact that AMOP protocol is less time-efficient protocol as it has the extensive voting phases in a round. Because they spend extra time in the voting phase, the estimated target utility in each bidding phase may be relatively lower than those in SAOP. That may be the reason the agents miss out on some good solutions for all parties. That also implies that there are less rounds within the same time period in AMOP compared to SAOP (3000 rounds Vs. 15000 rounds); therefore, there is less time to explore the outcome space. That is, 9000 offers were made during a negotiation in AMOP while agents made around between 22500 and 45000 offers in total in SAOP. As a future work, we would like test the protocols in a round-based deadline setting to see how their performance would be when they have the same number of rounds.

6 Discussion

The terms of *multiparty* and *multilateral* are used interchangeably in the community. In this work, we distinguish them as follows. If there are more than two participants engaged in the negotiation, it is considered a multiparty negotiation. This engagement can be in different forms such as *one-to-many*, *many-to-many* or *many-to-one* negotiations. For instance, William et al. propose a many-to-many concurrent negotiation protocol that allows agents to commit and to decommit their agreement [17]. Wong and Fang introduce the Extended Contract-Net-like multilateral Protocol (ECNPro) [1] for multiparty negotiations between a buyer and multiple sellers, which can be considered as multiple bilateral negotiations. In this work, we define multilateral negotiations as negotiations in which more than two agents negotiate in order to reach a joint agreement; in other words, all the negotiating parties have the same role during the negotiation process (e.g., a group of friends negotiating on their holiday), and these negotiations might or might not be mediated by an independent party that has no personal stake in the outcome of the negotiation.

The protocols proposed for multilateral negotiations in the multiagent community mostly use a mediator [2, 5, 9–11, 13]. In contrast, this paper proposes protocols for non-mediated multilateral negotiations. Endriss presents a monotonic concession protocol for non-mediated multilateral negotiations and discusses what a concession means in the context of multilateral negotiation, see [6]. The monotonic concession protocol enforces the agents to make a concession or to stick to their previous offer, while our protocols do not interfere with what to bid, only when to bid. The concession steps suggested in that work require to know the other agent's preferences except for the egocentric concession step in which the agent is expected to make a bid that is worse for itself.

A generalization of the alternating offers protocol, namely, a sequential-offer protocol was used in [18]. Similar to SAOP, the agents make sequential offers in predefined turns or accept the underlying offer according to this protocol. A minor difference is that it does not provide a walk-away option for the agents as SAOP does. The core of the work is a negotiation strategy that applies a sequential projection method for multilateral negotiations. In that sense it cannot be compared to the work presented in this paper, in which two multilateral negotiation protocols are proposed and evaluated.

De Jonge and Sierra recently introduced a new multilateral protocol inspired from human negotiations, called the *Unstructured Communication Protocol* (UCP) [4]. Unlike the negotiation protocols discussed above, this protocol does not structure the negotiation process. That is, any agent may propose an offer at any time and offers can be retracted at any time. Agents can accept a given offer by repeating the same offer. When all agents propose the same offer, this offer is considered an agreement. There are some similarities between their protocol and AMOP such as the agents can see multiple offers on the negotiation table and evaluate them. Compared to AMOP, their protocol is more flexible. For example, in AMOP agents have to bid in the bidding phase and have to vote in the voting phase, whereas agents in UCP can remain silent and wait for the other agents. However, flexibility comes with a price. Designing an agent having the intelligence to deal with the uncertainties in UCP is quite a challenge: how do you decide whether the agent should bid or remain silent? How do you know if another agent is still participating or whether it walked away? What does it mean if some of the agents are silent? Although the protocol is more natural from a human point of view, the situation is different: the agents lack information that humans that are physically present in the same negotiation room would have, such as body language, tone of voice, eye contact. Our point of view is that if we would like to develop a multilateral negotiation protocol in which humans and agents are to engage each other, then we should get the protocol as close as possible to the human way of negotiating, like UCP, while realizing that developing agents that can fully understand and act in such a heterogeneous setting is still a Grand Challenge. If, on the other hand, we are aiming for agents-only negotiations, then deviating from protocols that humans would use is quite alright, which opens the door for mechanism design, game theory and, of course, strategy development for the participating agents. The alternating multilateral negotiation protocols presented in this paper, are motivated by the search for protocols for agents-only negotiations.

7 Conclusion

In this paper, we introduce two extensions of alternating offers protocol for multilateral negotiations, namely *Stacked Alternating Offers Protocol* and *Alternating Multiple Offers Protocol*. We provide formal definitions of these protocols based on a general formalization for turn-taking multilateral negotiation protocols. Furthermore, we compare the performance of these protocols with time-based Conceder agents empirically. Our results show that SAOP performed better than AMOP in terms of social welfare and fairness of the negotiation outcome on the chosen negotiation scenarios. Therefore, we make SAOP public to facilitate the research in multilateral negotiation. In ANAC 2015 the participants developed negotiating agents for three-party negotiation governed by SAOP.

As future work, we are planning to characterize negotiation protocols using properties and show to which extent these are satisfied by SAOP, UCP, AMOP, and other (new) protocols. For instance, it would be interesting to investigate the effect of the agents' ordering in a given turn sequence on the negotiation outcome (e.g., whether or not the first agent starting the negotiation has an advantage over the others). In this work we used one type of agents in our evaluation of both protocols to ensure that we are not, at the same time, comparing negotiation strategies. However, it is still an open question what makes a protocol a good protocol. In future, we plan to perform more systematic evaluations for properties that characterize negotiation protocols, such as the speed with which agreements are reached, more fairness aspects (beyond distance to Nash Product, and ordering effects), scalability, robustness against manipulative agents (e.g., truth-revealing), and communication overhead.

Acknowledgements This work was supported by the ITEA M2MGrids Project, grant number ITEA141011.

References

1. A multi-agent protocol for multilateral negotiations in supply chain management. Int. J. Product. Res. **48** (2010)
2. R. Aydoğan, K.V. Hindriks, C.M. Jonker, Multilateral mediated negotiation protocols with feedback, in *Novel Insights in Agent-based Complex Automated Negotiation* (2014), pp. 43–59
3. T. Baarslag, K. Hindriks, C.M. Jonker, S. Kraus, R. Lin, The first automated negotiating agents competition (ANAC 2010), in *New Trends in Agent-based Complex Automated Negotiations*, Series of Studies in Computational Intelligence, ed. by T. Ito, M. Zhang, V. Robu, S. Fatima, T. Matsuo (Springer, Berlin, 2012), pp. 113–135
4. D. de Jonge, C. Sierra, Nb^3: a multilateral negotiation algorithm for large, non-linear agreement spaces with limited time. Auton. Agents Multi-Agent Syst. **29**(5), 896–942 (2015)
5. E. de la Hoz, M. Lopez-Carmona, M. Klein, I. Marsa-Maestre, Consensus policy based multi-agent negotiation, in *Agents in Principle*, Agents in Practice, volume 7047 of Lecture Notes in Computer Science, ed. by D. Kinny, J.-J. Hsu, G. Governatori, A. Ghose (Springer, Berlin, 2011), pp. 159–173

6. U. Endriss, Monotonic concession protocols for multilateral negotiation, in *Proceedings of the Fifth International Joint Conference on Autonomous Agents and Multiagent Systems*, Japan (2006), pp. 392–399
7. S. Fatima, S. Kraus, M. Wooldridge, *Principles of Automated Negotiation* (Cambridge University Press, New York, 2014)
8. S.S. Fatima, M. Wooldridge, N.R. Jennings, Optimal negotiation strategies for agents with incomplete information, in *Revised Papers from the 8th International Workshop on Intelligent Agents VIII*, ATAL '01, London, UK (Springer, Heidelberg, 2002), pp. 377–392
9. K. Fujita, T. Ito, M. Klein, Preliminary result on secure protocols for multiple issue negotiation problems, in *Proceedings of the Intelligent Agents and Multi-Agent Systems, 11th Pacific Rim International Conference on Multi-Agents, PRIMA 2008, Hanoi, Vietnam, 15–16 December, 2008*, (2008), pp. 161–172
10. H. Hattori, M. Klein, T. Ito, A multi-phase protocol for negotiation with interdependent issues, in *Proceedings of the 2007 IEEE/WIC/ACM International Conference on Intelligent Agent Technology, Silicon Valley, CA, USA, 2–5 November, 2007* (2007), pp. 153–159
11. M. Hemaissia, E. Seghrouchni, A., C. Labreuche, J. Mattioli, A multilateral multi-issue negotiation protocol, in *Proceedings of the Sixth International Joint Conference on Autonomous Agents and Multiagent Systems*, Hawaii (2007), pp. 939–946
12. H. Kameda, E. Altman, C. Touati, A. Legrand, Nash equilibrium based fairness. Math. Methods Oper. Res. **76**(1), 43–65 (2012)
13. M. Klein, P. Faratin, H. Sayama, Y. Bar-Yam, Protocols for negotiating complex contracts. IEEE Intell. Syst. **18**, 32–38 (2003)
14. L. Kleinrock, Analysis of a time shared processor. Nav. Res. Logist. Quart. **11**(1), 59–73 (1964)
15. R. Lin, S. Kraus, T. Baarslag, D. Tykhonov, K. Hindriks, C.M. Jonker, Genius: an integrated environment for supporting the design of generic automated negotiators. Comput. Intell. **30**(1), 48–70 (2014)
16. H. Raiffa, *The Art and Science of Negotiation: How to Resolve Conflicts and Get the Best Out of Bargaining* (Harvard University Press, Cambridge, 1982)
17. C.R. Williams, V. Robu, E.H. Gerding, N.R. Jennings, Negotiating concurrently with unknown opponents in complex, real-time domains, in *20th European Conference on Artificial Intelligence*, vol. 242, August 2012, pp. 834–839
18. R. Zheng, N. Chakraborty, T. Dai, K. Sycara, Multiagent negotiation on multiple issues with incomplete information: extended abstract, in *Proceedings of the 2013 International Conference on Autonomous Agents and Multi-agent Systems*, AAMAS '13 (2013), pp. 1279–1280

Atlas3: A Negotiating Agent Based on Expecting Lower Limit of Concession Function

Akiyuki Mori and Takayuki Ito

Abstract The sixth international Automated Negotiating Agents Competition (ANAC) was held in conjunction with the ninth international joint conference on Autonomous Agents and Multi-Agent Systems (AAMAS). We developed Atlas3 that is an automated negotiating agent for ANAC2015. In this paper, we explain about the searching methods and compromising strategy that is used by our agent. Our agent uses appropriate searching method based on relative utility for linear utility spaces. Moreover, our agent applies replacement method based on frequency of opponent's bidding history. Our agent decides concession value according to the concession function presented by us in Mori and Ito (A compromising strategy based on expected utility of evolutionary stable strategy in bilateral closed bargaining problem, 2015, [2]). In Mori and Ito (A compromising strategy based on expected utility of evolutionary stable strategy in bilateral closed bargaining problem, 2015, [2]), we derived an estimated expected utility to estimate an appropriate lower limits of concession function. However, Mori and Ito (A compromising strategy based on expected utility of evolutionary stable strategy in bilateral closed bargaining problem, 2015, [2]) proposes a concession function for bilateral multi-issue closed bargaining games. Therefore, we extend the concession function for multi-lateral multi-issue closed bargaining games.

Keywords Automated multi-issue negotiation · Compromising strategy · Automated negotiating agents competition

---

A. Mori (✉) · T. Ito
Nagoya Institute of Technology, Aichi, Japan
e-mail: mori.akiyuki@itolab.nitech.ac.jp

T. Ito
e-mail: ito.takayuki@nitech.ac.jp

© Springer International Publishing AG 2017

K. Fujita et al. (eds.), *Modern Approaches to Agent-based Complex Automated Negotiation*, Studies in Computational Intelligence 674,
DOI 10.1007/978-3-319-51563-2_11

1 Introduction

The sixth international Automated Negotiating Agents Competition (ANAC) was held in conjunction with the ninth international joint conference on Autonomous Agents and Multi-Agent Systems (AAMAS) on May 2015. Researchers from the worldwide negotiation community participate in ANAC [1].

We propose an automated negotiating agent Atlas3 that is developed for ANAC2015. Our agent uses an relative utility search for linear utility spaces. In this paper, relative utility is based on the maximum utility as a standard. Agent can get a bid (an agreement candidate) that satisfy conditions. Moreover, our agent apply replacement method based on frequency of opponent's bidding history. Our agent uses compromising strategy presented by us in [2]. The method is a compromising strategy of bilateral multi-issue closed bargaining games. Therefore, we extend the method for multi-lateral multi-issue closed bargaining games. In [2], we analyze a final phase of bargaining game as strategic form games. Then, we derive estimated expected utility in a equilibrium point of evolutionarily stable strategies [3]. Finally, we set the estimated expected utility as a lower limit of concession function.

2 Searching Methods

Our agent uses the relative utility search and the replacement method based on frequency of opponent's bidding history. First, we explain about the relative utility search. Our agent derives a relative utility matrix according to our agent's utility space. Figure 1 shows an example of relative utility matrix. In Fig. 1, there are three issues i_1, i_2, and i_3 and i_n has three values v_{n1}, v_{n2}, and v_{n3}. Value which is enclosed in parentheses means relative utility of v_{nk} in Fig. 1. For example, relative utility of v_{12} is -0.2. Our agent derives relative utility based on the maximum utility. In Fig. 1, when a bid has three values v_{11}, v_{12}, and v_{31}, utility of the bid is 1.0 which is maximum utility. Therefore, relative utility of a bid which has three values v_{11}, v_{21} is -0.1 in linear utility spaces because utility of the bid is 0.9 ($0.9 - 1.0 = -0.1$). The relative utility search searches bids that satisfy a concession function based on maximum utility bid. For example, When a bid has two values v_{12}, v_{21} and threshold

i_1	i_2	i_3
v_{11} (0.0)	v_{21} (0.0)	v_{31} (**0.0**)
v_{12} (- 0.1)	v_{22} (- 0.5)	v_{32} (**- 0.2**)
v_{13} (- 0.2)	v_{23} (- 0.2)	v_{33} (- 0.3)

Fig. 1 Relative utility matrix threshold value = 0.7

i_1	i_2	i_3
v_{11} (10 times)	v_{21} **(35 times)**	v_{31} (5 times)
v_{12} (20 times)	v_{22} (10 times)	v_{32} (0 times)
v_{13} **(30 times)**	v_{23} (15 times)	v_{33} **(55 times)**

Fig. 2 Frequency matrix

value based on concession function is 0.7, our agent can offer bids which has utility more than threshold value 0.7. Consequently, our agent can selects v_{31} or v_{32} in i_3.

We adopt the relative utility search based on a maximum utility bid. Our agent randomly sorts issues and replaces value of each issue in order. If a bid which is replaced values satisfy conditions, our agent adopts the bid. By means of this method, computational complexity of the method is $O(n)$, where n means the number of issues.

Next, we explain about the replacement method based on frequency of opponent's bidding history. Our agent updates a frequency matrix that has the number of occurrences of values each round. Figure 2 shows an example of frequency matrix. Value which is enclosed in parentheses means the number of occurrences of v_{nk} in Fig. 2. For example, the number of occurrences of bids which include v_{nk} is 10. Our agent replaces a bid that is searched by the relative utility search based on a frequency matrix.

A value frequently appearing on each issue is the subject of replacement. In Fig. 2, $v_{13}(30times)$, $v_{21}(35times)$ and $v_{33}(55times)$ are appropriate. Our agent adopts the replaced bid if the replaced bid has utility more than threshold value.

3 Expecting Lower Limit of Concession Function

In [2], we proposed a compromising strategy for bilateral multi-issue closed bargaining games. We extend the compromising strategy for multi-lateral multi-issue closed bargaining games. To determine the appropriate concession value in bargaining games, we divide the negotiating flow into two phases. Figure 3 shows the negotiating flow and the divided phases. We define an alternating offers phase (AOP) and a final offer phase (FOP). Our agent decides AOP's concession value by FOP's expected utility. We explain about changes of [2]. We regarded FOP as strategic form games in [2]. Moreover, we derived estimated expected utility of equilibrium point of evolutionarily stable strategies. In this paper, we derive the estimated expected utility for multi-lateral multi-issue closed bargaining games.

The number of opponent is only one in bilateral multi-issue closed bargaining games. On the other hand, the number of opponents is more than one in multi-lateral multi-issue closed bargaining games. Therefore, our agent regards a bid that has the maximum utility in bids are accepted by opponents as a compromise bid.

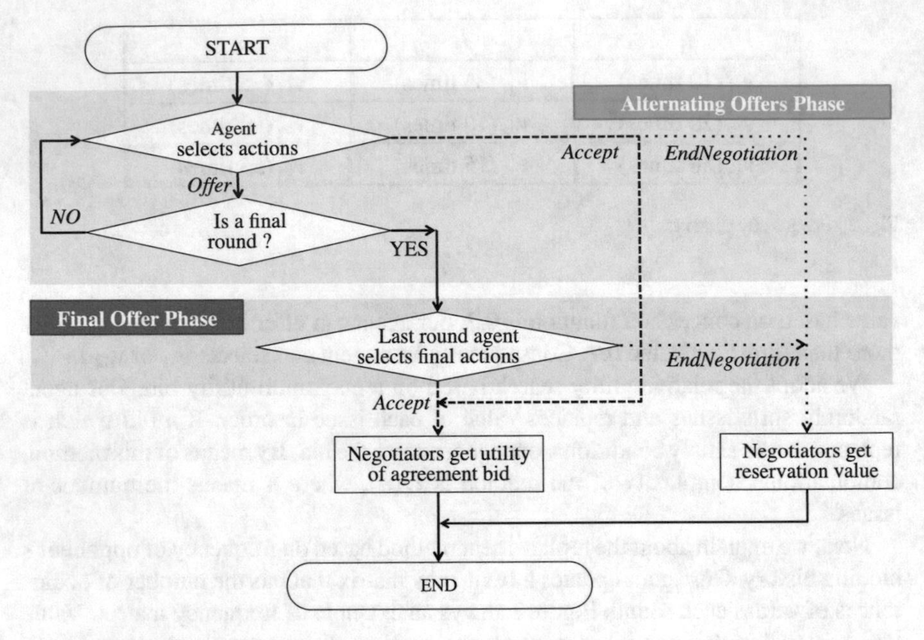

Fig. 3 Phases of negotiation flow

In addition, the utility of compromise bid is reservation value if any bids are not accepted by opponents.

Our agent regards the estimated expected utility in consideration of discounted utility as a lower limit of concession function $L(t)$, where t means normalized time $(0.0 \leq t \leq 1.0)$. Our agent's concession function $T(t)$ is designed as formula (1).

$$T(t) = \begin{cases} L(t) + (1.0 - L(t)) \cdot (1.0 - t) & (df_A = 1.0) \\ 1.0 - t/\alpha & (df_A < 1.0 \wedge 1.0 - t/\alpha \geq L(t)) \\ L(t) & (df_A < 1.0 \wedge 1.0 - t/\alpha < L(t)) \end{cases} \quad (1)$$

df means a discount factor in the formula (1) and Atlas3 sets $\alpha = df$ based on heuristics. Figure 4 shows $T(t)$ in cases 1, 2, 3, and 4.

case 1: $df = 1.0 \text{C} RV = 0.00 \text{C} C = 0.5$
case 2: $df = 0.5 \text{C} RV = 0.00 \text{C} C = 0.5$
case 3: $df = 1.0 \text{C} RV = 0.75 \text{C} C = 0.5$
case 4: $df = 0.5 \text{C} RV = 0.75 \text{C} C = 0.5$

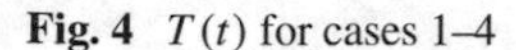

Fig. 4 $T(t)$ for cases 1–4

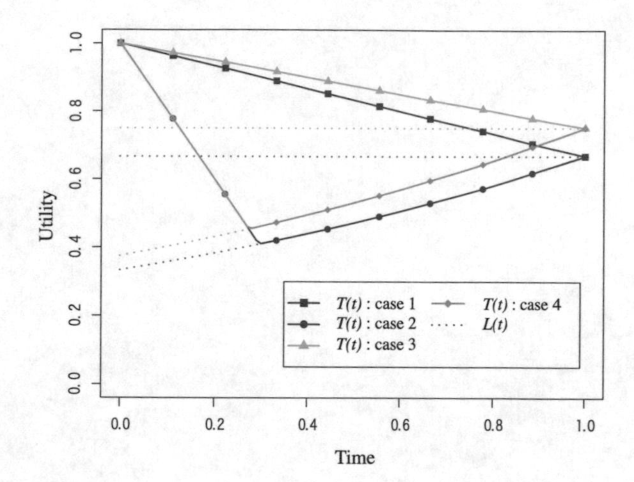

4 Conclusion

In this paper, we have proposed the Atlas3 that is an automated negotiating agent for ANAC2015. We have explained about the relative utility search and the replacement method based on frequency of opponent's bidding history. Moreover, we have mentioned a negotiation strategy of Atlas3 that is based on the estimated expected utility.

References

1. T. Baarslag, K. Hindriks, C.M. Jonker, S. Kraus, R. Lin, The first automated negotiating agents competition (ANAC 2010), *New Trends in Agent-Based Complex Automated Negotiations*, Series of Studies in Computational Intelligence (Springer, Berlin, 2012), pp. 113–135
2. A. Mori, T. Ito, A compromising strategy based on expected utility of evolutionary stable strategy in bilateral closed bargaining problem, in *Proceedings of Agent-Based Complex Automated Negotiations* (2015), pp. 58–65
3. J.M. Smith, *Evolution and the Theory of Games* (Cambridge University Press, Cambridge, 1982)

Pars Agent: Hybrid Time-Dependent, Random and Frequency-Based Bidding and Acceptance Strategies in Multilateral Negotiations

Zahra Khosravimehr and Faria Nassiri-Mofakham

Abstract We propose our Pars agent, one of the ANAC 2015 winners in multilateral negotiation tournaments. In this challenge, any agreement is made through acceptances issued by all parties involved in a trilateral negotiation. Pars agent uses a hybrid bidding strategy that combines behaviors of time-dependent, random and frequency-based strategies to propose a high utility offer close to the opponents bids and to increase the possibility of early agreement.

1 Introduction

In this study, we propose Pars agent and its strategy which was designed to participate in Automated Negotiating Agents Competition [1–6] and could finish at 2nd rank in the individual utility category in the sixth ANAC [6]. In ANAC2015, a set of automated negotiating agents with linear utility functions compete against each other through several rounds of multilateral alternating offers using SAOP (Stacked Alternating Offers Protocol), a simple extension to bilateral alternating offers protocol [7], in GENIUS environment [8]. According to SAOP, all participants in session get a turn per round. The first party starts with an offer that is immediately observed by all others. After observing the offer, the next party can make a counter-offer (i.e., rejecting the proposal), accept the offer, or quit without any agreement. This alternating offers is repeated until an agreement or the deadline is reached. To reach an agreement, all parties should accept the offer before the deadline. Otherwise, the negotiation fails [7].

In accordance to BOA (bidding strategy, opponent model, and acceptance strategy) architecture [9], Pars agent employs a simple opponent modeling by just

Z. Khosravimehr · F. Nassiri-Mofakham (✉)
Department of Information Technology Engineering, University of Isfahan, Isfahan, Iran
e-mail: fnasiri@eng.ui.ac.ir

Z. Khosravimehr
e-mail: z.khosravimehr@eng.ui.ac.ir

© Springer International Publishing AG 2017

K. Fujita et al. (eds.), *Modern Approaches to Agent-based Complex Automated Negotiation*, Studies in Computational Intelligence 674,
DOI 10.1007/978-3-319-51563-2_12

considering the previous offers proposed by the other two parties. However, for bidding and accepting offers follows a hybrid time-dependent, random and frequency-based method.

The rest of the paper is organized as follows. In Sects. 2 and 3, bidding and acceptance strategies of Pars agent on how to bid and when to accept are proposed, in which opponent modeling is embedded as keeping a copy of the bids previously offered by the other parties. Section 4 shows the results and Sect. 5 concludes the paper.

2 Acceptance Strategy

Pars agent uses time-dependent strategy for both accepting a bid or making a new offer. In time-dependent strategy [10], a real number G_T, called target utility, is computed as Eq. 1,

$$G_T = u_T(t) = 1 - t^{\frac{1}{\beta}} \quad (1)$$

in which, u_T is a time dependent utility function to compute the target utility, t and β are time and the value to control the concession speed of the agent, respectively [11]. β equals to, greater than, and less than 1 means Fixed Concession, Conceder, and Boulware behavior, respectively [12]. Pars agent behavior is Boulware. It considers $\beta = 0.2$ for discounted domains and $\beta = 0.15$ for the domains without any discount factors. Based on several experiments achieved in Genius, we set these values as such aiming the agent faces less number of failed negotiations and gains appropriate utilities against different opponents. In discounted domains, the utility of an offer is decreased as the negotiation time passes, and so the faster the agent compromises, the less benefit it will lose. Whenever a new offer is received, Pars agent computes its target utility. If G_T is greater than 0.7, Pars agent behaves Hardheaded and does not accept any offer with the utility[1] below G_T. However, when G_T is less than 0.7, the agent considers G_T as constant value 0.7, behaves constant, and accepts any offer with the utility above 0.7. Figure 1 shows the utility diagram of Pars agent in a domain without discount factor.

3 Bidding Strategy

With regard to bidding, Pars agent adds randomness to its behavior. In the first round of a negotiation, if Pars agent be the first mover, it chooses its best bid, that is (i.e., the bid which has the highest utility[2] for Pars agent). But, due to its random

[1] The utility can be a weighted sum of the utilities associated with the values of each issue [13, 14].

[2] The utility can be a weighted sum of the utilities associated with the values of each issue [11, 13, 14, 16]

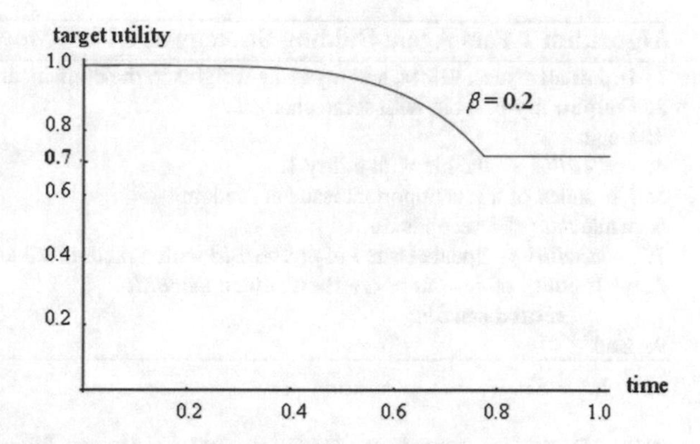

Fig. 1 The Target utility of Pars agent for discounted domains

Table 1 A sample bid

Food	Drinks	Location	Invitations	Music	Cleanup
Chips and nuts	Beer only	Party room	Photo	DJ	Specialized materials

behavior and avoiding being exploited by other parties, Pars Agent changes this bid and then proposes this offer. So, it exactly doesn't start with its best bid, which is the dominant approach in the literature [17]. Algorithm 1 shows this strategy. However, if the other parties move first or after the first round, Pars agent also considers the offers proposed by the other parties in previous rounds and follows random and frequency-based methods as shown in Algorithms 2 and 3.

3.1 Bidding Strategy 1: Pars Agent Moves First

Each bid comprises a set of values of the issues representing features of a negotiation domain. For example, Table 1 shows the bid for the agent with values Chips and Nuts, Beer Only, Party Room, Photo, DJ, and Specialized Materials respectively for issues Food, Drinks, Location, Invitations, Music, and Cleanup in a domain.

Inspired by some idea behind WALKSAT algorithm [15] and adapting it to the problem, Pars agent chooses one of the issues in its best bid at random. The higher the issue weight, the less it is chosen. It then changes the value of this issue to one of the issue values at random. If the utility of the new bid is greater than the current target utility G_T of the agent, it is proposed. Otherwise, Pars agent repeats changing the issue value until observing the new bid yields a suitable benefit or passing 3 s.[3] In the later case, the agent proposes the last generated bid. Since the initial bid has the highest utility, could bring the highest utility for the agent, it is high probable that the proposed bid brings the agent a utility above its current target utility G_T.

[3]Each negotiation lasts 3 minutes [8].

Algorithm 1 Pars Agent Bidding Strategy as a First Mover

1: **Input:** all issues, values, and my issue weights in the domain, and my current target utility G_T.
2: **Output:** my best bid with some changes.
3: **Begin**
4: *myBestBid* ← the bid with utility 1.
5: i ← index of a less important issue at random.
6: **while** *time* ≤ 3 seconds **do**
7: *newBid* ← update issue i of *myBestBid* with a random value;
8: **if** utility of *newBid* > G_T **then return** *newBid*;
return *newBid*;
9: **End**

3.2 *Bidding Strategy 2: The Other Party Moves First*

When Pars agent is not a first mover, then its bid depends on the bids the other parties proposed before. If the benefits of the latest proposal is higher than its current target utility, it accepts the proposal and does not bid a new offer. Otherwise, Pars follows two steps.

For proposing a bid, it first takes into account the list of proposals mutually accepted by both opponent agents. Per each such a bid, Pars agent chooses one of the issues at random. The higher the issue weight for Pars agent, the more it is chosen. It then assigns a value to this issue at random. If the utility of the new bid is greater than the current target utility G_T of the agent, it is proposed. Otherwise, Pars agent repeats this step for all other proposals in the list to see a bid whose utility is above G_T. This step of the second bidding strategy of Pars agent is illustrated in Algorithm 2.

Algorithm 2 Pars Agent Bidding Strategy in Next Rounds – Step 1

1: **Input:** all issues, values, and my issue weights in the domain, my current target utility G_T, and list L of proposals mutually accepted by opponent agents.
2: **Output:** my adapted best bid among mutually accepted proposals of the opponents.
3: **Begin**
4: *maxBid* ← null.
5: **for** s 1 **to** number of issues in the domain **do**
6: **while** any proposal P exists in the list L **do**
7: i ← index of a more important issue of P at random;
8: *newBid* ← update issue i of P with a random value;
9: **if** utility of *newBid* > $max\{G_T, utility\ of\ maxBid\}$ **then**
10: *maxbid* ← *newBid*;
11: **return** *maxBid*;
12: **End**

If no bid generated from the first step, Pars agent uses a frequency-based strategy to look for overlaps in proposals made by other agents. It considers the frequency of the values each opponent proposed for any issue and in descending order. Tables 2

Table 2 Frequency of the values proposed by opponent A

Attribute 1: Food		Attribute 3: Drinks		Attribute 3: Invitations	...
Frequency	Value	Frequency	Value	Frequency	Value
3	Chips and nuts	4	Beer only	4	Party room
2	Catering	2	Non-alcoholic	2	Party tent
1	Finger-food			1	Ballroom

Table 3 Frequency of the values proposed by opponent B

Attribute 1: Food		Attribute 3: Drinks		Attribute 3: Invitations	...
Frequency	Value	Frequency	Value	Frequency	Value
3	Chips and nuts	3	Non-alcoholic	3	Party room
2	Catering	2	Handmade cocktails	2	Ballroom
		1	Beer only		

and 3 show sample list of the values proposed by opponents A and B along with their frequencies each in descending order.

For each issue, if the values (corresponding the highest frequency) proposed by each opponent are equal, Pars agent also uses this value for the issue in generating its new bid. For example, it sets the value of issue Food as Chips and Nuts since it is the high frequency value proposed by two agents A and B for this issue.

If the number of issues with no assigned values is greater than half of the total number of the issues in the domain, Pars agent restarts this step for all issues but by looking for the first and the second highest common value proposed by the opponents. For example, agent A proposed values Beer Only and Non-Alcoholic for issue Drink 4 and 2 times, respectively. However, agent B offered values Non-Alcoholic and Handmade Cocktails for the same issue 3 and 2 times, respectively. Then, Pars agent assigns Non-Alcoholic to issue Drink in its new bid. By doing so, it tries to generate bids which are more favorable and beneficial for its opponents. This decreases the number of empty issues. For the issues which remained empty, Pars agent assigns its own most favorite values. Algorithm 3 shows the second step Pars agent follows in its second bidding strategy. Finally, if none of the described methods lead to a proposal with enough utility, Pars agent propose its best bid with some changes similar to what described in Algorithm 1. The result of employing the algorithm on Tables 2 and 3 is shown in Table 4.

Algorithm 3 Pars Agent Bidding Strategy in Next Rounds – Step 2

```
1: Input: all issues, values, and my issue weights in the domain, and lists L_A and L_B of proposals
      by agents A and B, respectively.
2: Output: my best bid extracted from L_A and L_B.
3: Begin
4: cycle ← 2.
5: d ← number of issues in the domain
6: while cycle > 0 do
7:     for i ← 1 to d do
8:         v_A ← max value of issue(i) in L_A;
9:         v_B ← max value of issue(i) in L_B;
10:        if cycle == 2 then
11:            if v_A == v_B then
12:                newBid.issue(i) ← v_A;
13:            else
14:                newBid.issue(i) ← null.
15:        else
16:            if v_A == v_B then
17:                newBid.issue(i) ← v_A;
18:            else
19:                sv_B ← second max value of issue(i) in L_B;
20:                if v_A == sv_B then
21:                    newBid.issue(i) ← v_A;
22:                else
23:                    sv_A ← second max value of issue(i) in L_A;
24:                    if sv_A == v_B then
25:                        newBid.issue(i) ← v_B;
26:                    else
27:                        if sv_A == sv_B then
28:                            newBid.issue(i) ← sv_B;
29:                        else
30:                            newBid.issue(i) ← null;
31:        if cycle != 0 then
32:            cycle ← cycle - 1;
33:            n ← number of null issues in newBid;
34:            if n / d ≥ 0.5 then
35:                cycle ← cycle - 1;
36:        else
37:            cycle ← cycle - 1;
38:        Forall null issues i;
39:        newBid.issue(i) ← my most favorite value for issue i;
40:        return newBid;
41: End
```

Table 4 The issue values of newBid based on highest common values proposed by the opponents

Step	Food	Drinks	Location	Invitations	Music	Cleanup
1:	Chips and nuts	null	Party room	Null	Null	Null
1':	Chips and nuts	**Non-alcoholic**	Party room	Null	Null	Null
			…			
2:	Chips and nuts	Non-alcoholic	Party room	**Custom**	**Band**	**Hired help**

4 Results

GENIUS framework for ANAC2015 provides several sample opponents. Figure 2 shows the results of experimenting Pars agent against these opponents in a multi-party negotiation session where Party1, Party2, and Party3 are Pars, a Boulware, and a Conceder agents, respectively. Table 5 illustrates the results observed in Fig. 2 which yielded a trilateral agreement in 2.178 s.

According to ANAC2015 experiments, Pars agent reached the final in Individual Utility category and achieved individual utility 0.47 in average in the final tournaments. Table 6 summarizes the results [6].

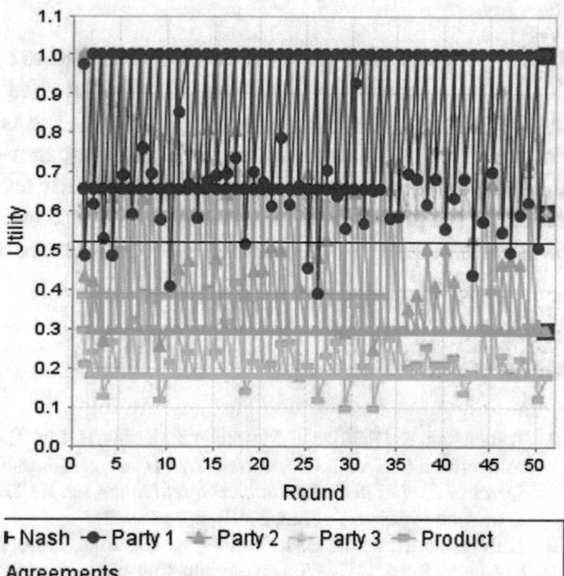

Fig. 2 Result of competition between Pars and Boulware and Conceder agents (*Party1* Pars agent, *Party2* a Boulware agent, and *Party3* a Conceder agent)

Table 5 Numerical result of competition between Pars and Boulware and Conceder agents (Party1: Pars agent, Party2: a Boulware agent, and Party3: a Conceder agent)

Time (s)	Rounds	Agreement?	Discounted?	Approval	Min. utility	Max. utility
2.17845935	51	Yes	No	3	0.29693	1.00000
Distance to pareto	Distance to Nash	Social welfare	Agent utility (Party 1)	Agent utility (Party 2)	Agent utility (Party 3)	
0.00000	0.67808	1.89378	1.00000	0.29693	0.59685	

Table 6 ANAC15 final results in individual utility category [6]

Agent name	Average	Standard deviation
Atlas3	0.481042722	0.00156024
ParsAgent	**0.470693979**	**0.003128712**
RandomDance	0.46062548	0.003038258
kawaii	0.460129481	0.002715924
agentBuyog	0.458823101	0.003842303
PhoenixParty	0.442975836	0.005032221
XianFaAgent	0.353133027	0.001918821
PokerFace	0.344003999	0.001433044

5 Conclusion

This paper proposed Pars agent that was designed to participate in ANAC 2015. The 2015 tournament is multilateral in which each agent negotiates with three agents in each session. They reach an agreement, if a bid is accepted by all three agents. Pars agent follows a mixture of random and frequency-based bidding and acceptance strategies. It also considers its current target utility and the utility of received bids in generating or proposing offers. Pars agent finished at the second place in the category of individual utility in the final ANAC2015 tournaments.

References

1. T. Baarslag, K. Hindriks, C.M. Jonker, S. Kraus, R. Lin, The first automated negotiating agents competition (ANAC 2010), in *New Trends in Agent-based Complex Automated Negotiations, Series of Studies in Computational Intelligence*, ed. by T. Ito, M. Zhang, V. Robu, S. Fatima, T. Matsuo (Springer, Berlin, 2012), pp. 113–135
2. T. Baarslag, K. Fujita, E.H. Gerding, K. Hindriks, T. Ito, N.R. Jennings, C. Jonker, S. Kraus, R. Lin, V. Robu, C.R. Williams, Evaluating practical negotiating agents: Results and analysis of the 2011 international competition. Artif. Intell. **198**, 73–103 (2013)
3. C.R. Williams, V. Robu, E.H. Gerding, N.R. Jennings, An overview of the results and insights from the third automated negotiating agents competition (ANAC2012), in *Novel Insights in*

Agent-based Complex Automated Negotiation, ed. by I. Marsá-Maestre, M.A. Lápez-Carmona, T. Ito, M. Zhang, Q. Bai, K. Fujita (Springer, Berlin, 2012), pp. 151–162
4. Kobi (Ya'akov) Gal, L. Ilany, The fourth automated negotiation competition, in *Next Frontier in Agent-based Complex Automated Negotiation*, ed. by K. Fujita, T. Ito, M. Zhang, V. Robu (Springer, Berlin, 2013), pp. 129–136
5. The Fifth Automated Negotiating Agents Competition (ANAC 2014). (2014), http://www.itolab.nitech.ac.jp/ANAC2014/. Accessed 31 Jan 2015
6. The Sixth Automated Negotiating Agents Competition (ANAC 2015). (2015), http://www.tuat.ac.jp/~katfuji/ANAC2015/. Accessed 31 Jan 2015
7. R. Aydoğan, D. Festen, K.V. Hindriks, C.M. Jonker, Alternating offers protocols for multilateral negotiation, in *Modern Approaches to Agent-Based Complex Automated Negotiation*, ed. by K. Fujita, Q. Bai, T. Ito, M. Zhang, R. Hadfi, F. Ren, R. Aydoğan (Springer, To be published)
8. R. Lin, S. Kraus, T. Baarslag, D. Tykhonov, K. Hindriks, C.M. Jonker, Genius: an integrated environment for supporting the design of generic automated negotiators. Comput. Intell. **30**(1), 48–70 (2014)
9. T. Baarslag, K. Hindriks, M. Hendrikx, A. Dirkzwager, C. Jonker, Decoupling negotiating agents to explore the space of negotiation strategies, in *Novel Insights in Agent-based Complex Automated Negotiation*, vol. 535. Studies in Computational Intelligence, ed. by I. Marsa-Maestre, M.A. Lopez-Carmona, T. Ito, M. Zhang, Q. Bai, K. Fujita (Springer, Berlin, 2014), pp. 61–83
10. P. Faratin, C. Sierra, N.R. Jennings, Negotiation decision functions for autonomous agents. Robot. Auton. Syst. **24**(3), 159–182 (1998)
11. F. Zafari, F. Nassiri-Mofakham, Bravecat: iterative deepening distance-based opponent modeling and hybrid bidding in nonlinear ultra large bilateral multi issue negotiation domains, in *Recent Advances in Agent-based Complex Automated Negotiation*, ed. by T. Ito, K. Fujita, R. Hadfi, T. Baarslag. Studies in Computational Intelligence (Springer International Publishing), pp. 285–293 (2016)
12. T. Baarslag, *What to Bid and When to Stop*. TU Delft, Delft University of Technology, 2014
13. F. Nassiri-Mofakham, M.A. Nematbakhsh, N. Ghasem-Aghaee, A. Baraani-Dastjerdi, A heuristic personality-based bilateral multi-issue bargaining model in electronic commerce. Int. J. Hum.-Comput. Stud. **67**(1), 1–35 (2009)
14. Farhad Zafari, Faria Nassiri-Mofakham, and Ali Zeinal Hamadani. Dopponent: A socially efficient preference model of opponent in bilateral multi issue negotiations. J. Comput. Secur. **1**(4), 283–292 (2015)
15. S. Russell, P. Norvig, Artificial intelligence: a modern approach (2010), Pearson
16. F. Nassiri-Mofakham, M.A. Nematbakhsh, A. Baraani-Dastjerdi, N. Ghasem-Aghaee, R. Kowalczyk, Bidding strategy for agents in multi-attribute combinatorial double auction. Expert. Syst. Appl. **42**(6), 3268–3295 (2015), Elsevier
17. F. Zafari, F. Nassiri-Mofakham, POPPONENT: Highly accurate, individually and socially efficient opponent preference model in bilateral multi issue negotiations. Artif. Intell. **237**, 59–91 (2016)

RandomDance: Compromising Strategy Considering Interdependencies of Issues with Randomness

Shinji Kakimoto and Katsuhide Fujita

Abstract In multi-lateral negotiations, agents need to simultaneously estimate the utility functions of more than two agents. In this chapter, we propose an estimating method that uses simple weighted functions by counting the opponent's evaluation value for each issue. For multi-lateral negotiations, our agent considers some utility functions as the 'single' utility function by weighted-summing them. Our agent needs to judge which weighted function is effective and the types of the opponents. However, they depend on the domains, agents' strategies and so on. Our agent selects the weighted function and opponent's weighting, randomly.

1 Estimating Utility Functions by Counting Values

In SAOP [1], the opponent's bids proposing many times are important. However, it is hard to get the statistical information by simply counting all of them because the proposed bids are limited in one-shot negotiations. Therefore, we propose a novel strategy that estimates the utility functions by counting the value of the opponent's bids in multi-lateral negotiations.

In our definitions, A_0 is our agent and $a(a = \{A_1, A_2\})$ are two opponents among the $three$-lateral negotiations. Agent a's previous bids are represented as B_a. The estimated utility function of agent a is represented as $eval'_a()$, which is defined as Eq. 1:

$$eval'_a(\mathbf{s_i}) = \sum_{\mathbf{s}' \in B_a} boolean(\mathbf{s}_i, \mathbf{s}') \cdot \mathbf{w(s')}. \quad (1)$$

S. Kakimoto (✉) · K. Fujita
Faculty of Engineering, Tokyo University of Agriculture and Technology, Tokyo, Japan
e-mail: kakimoto@katfuji.lab.tuat.ac.jp

K. Fujita
e-mail: katfuji@cc.tuat.ac.jp

© Springer International Publishing AG 2017

K. Fujita et al. (eds.), *Modern Approaches to Agent-based Complex Automated Negotiation*, Studies in Computational Intelligence 674,
DOI 10.1007/978-3-319-51563-2_13

The function $boolean(\mathbf{s}_i, \mathbf{s}')$ returns 1 when bid s' contains the s_i, and otherwise it returns 0. Function $w(\mathbf{s})$ is the weighting function that reflects the order of the proposed bids. Therefore, estimated utility function $U'_a(\mathbf{s})$ of Alternative solutions: $\mathbf{s}$ of opponent a is defined as Eq. 2:

$$U_a(\mathbf{s}) = \frac{u_a(\mathbf{s})}{\max_{\forall \mathbf{s}'} u_a(s')} \tag{2}$$

$$u_a(\mathbf{s}) = \sum_{i=1}^{N} eval'_a(s_i). \tag{3}$$

Using Eq. 2, our agent can obtain estimated utility that is normalized [0, 1] to each opponent.

In addition, our agent selects a weighting function from the following ones randomly:

- Constant Function: $w(\mathbf{s}) = 1$
- Exponential Growth Function: $w(\mathbf{s}) = 1.05^{count_a(\mathbf{s})}$
- Exponential Decay Function: $w(\mathbf{s}) = 0.95^{count_a(\mathbf{s})}$

$count_a(\mathbf{s})$ is a function that returns the number of previous bid's when agent a proposes bid $\mathbf{s}$.

Figure 1 is the examples of three weighting functions. Three kinds of weighting functions are introduced to our agent: *Constant Function, Exponential Growth Function, Exponential Decay Function. Constant Function* is the same weight when the time passes. However, *Exponential Growth Function* and *Exponential Decay Function* change as the time passes. They are effective when the opponent's strategy can be estimated, however, it is often hard to judge the opponent's types of strategy with accuracy.

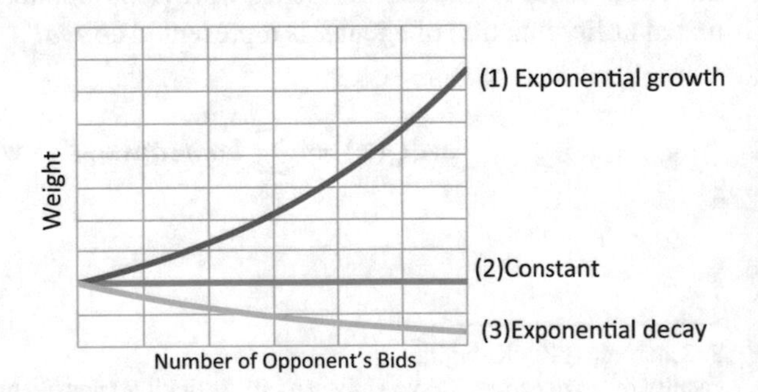

Fig. 1 Examples of three weighting functions

2 Weighted Sum of the Estimated Utility

We proposed an agent's strategy using the estimated utility function in Sect. 1. In multi-lateral negotiations, a novel strategy needs to determine how much our agent can compromise to each agent using the estimated utility functions. Our proposed agent employs $h_n (n = 1, 2)$, which is a function for compromising to each agent(A_1, A_2), to judge subsequent bids. The evaluation function (U_{op}) of the bid (**s**) that combines two opponents is defined as Eq. 4:

$$U_{op}(\mathbf{s}) = \sum_{n=1}^{2} h_n U_{A_n}(\mathbf{s}). \tag{4}$$

Our agent can adopt a negotiation strategy for bilateral negotiations to multi-lateral negotiations by combining the utility functions of opponents.

In addition, our agent uses the following four weighting functions for compromising to each agent:

- Same weighting: $(h_1, h_2) = (0.5, 0.5)$
- Consider only A_1: $(h_1, h_2) = (1, 0)$
- Consider only A_2: $(h_1, h_2) = (0, 1)$
- Cooperative weighting.

When our agent selects the same weighting, it decides the U_{op} by considering the both agents equally. When our agent selects the weighting considering only A_1, it decides the U_{op} by considering the A_1, only. When our agent selects the weighting considering only A_2, it decides the U_{op} by considering the A_2, only. Cooperative weighting is defined as the ratio the opponent proposes the bids with high social welfare. Agent a's k-th previous bid are represented as $\mathbf{s}_k^a$. h_1 and h_2 is calculated by Algorithm 5. h_a is calculated as the count of bids which is the best estimated social welfare in the recent N rounds. As the agent a offers the bid with high social welfares, h_a increases. In ANAC2015, $N = 200$ by tuning the parameters.

Algorithm 5 Calculate h_n in the cooperative weighting

```
h1 ⇐ 0, h2 ⇐ 0
i ⇐ 0
while i < N do
    if SocialWelfare(s_i^{n1}) > SocialWelfare(s_i^{n2}) then
        h1 ⇐ h1 + 1
    else
        h2 ⇐ h2 + 1
```

3 Strategy of RandomDance

We propose an agent's strategy using the estimated utility function in Sect. 1 and the weighted sum of the estimated utility in Sect. 2. When a new negotiation starts, our agent proposes three estimated utility functions to each opponent's using weighting functions in Sect. 1. In choosing the action, our agent selects the estimated utility function to each opponent and the compromising function in Sect. 2, randomly. By selecting the function from some of the weighting functions randomly, our agent can judge to some types of opponent's despite that the opponent has rare strategies. In addition, the randomness has sometimes effective in the unconfirmed situations. Our agent decides its next action based on Eqs. 5 and 6:

$$target_{end} = U_{my}(arg\ max\ EstimatedSocialWelfare) \tag{5}$$

$$target(t) = \begin{cases} (1 - t^3)(1 - target_{end}) + target_{end} & (d = 1) \\ (1 - t^d)(1 - target_{end}) + target_{end} & (otherwise) \end{cases} \tag{6}$$

$Target_{end}$ is defined as our utility of bid with the best social welfare. Our agent proposes a bid whose utility exceeds $target(t)$ and the highest U_{op} (Eq. 4). Figure 2 shows the changes of $target(t)$ when $target_{end} = 0.5$.

When $d = 1$ the discount is none in the negotiation, our agent compromises slowly. When the discount factor d is small, our agent compromises rapidly by considering the conflicts among agents based on $target_{end}$. As the discount factor d becomes smaller, the agent compromises in the earlier stage. By deciding the proposal and acceptance strategies using Eq. 5, the agent can compromise considering the rate of conflicting. It accepts the opponent's bids when they are more than $target(t)$. In addition, our agent terminates this negotiation when it can't gain more than the reservation value; in other words, $target(t)$ is less than the reservation value.

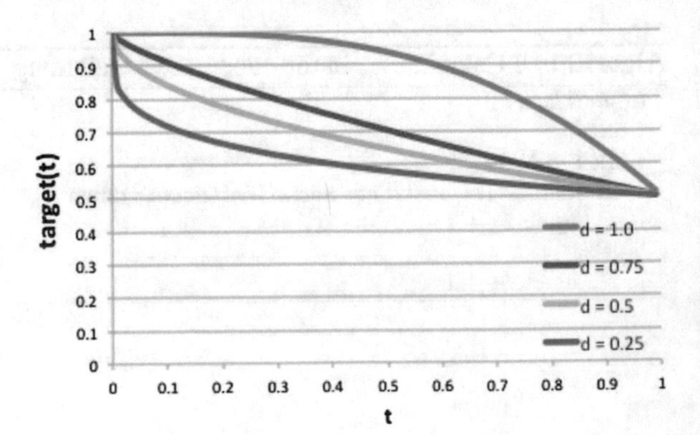

Fig. 2 Example of threshold of offer and accept ($target_{end} = 0.5$)

4 Conclusion

We proposed an estimating method that uses simple weighted functions by counting the opponent's evaluation value for each issue. For multi-lateral negotiations, our agent considered some utility functions as the 'single' utility function by weighted-summing them. Our agent needed to judge the effective weighted function and the opponent. However, the weighted function depended on the situations. Our agent selected the effective weighted function and opponent's considering weights, randomly.

Future works will address improvements in estimating the opponent's utility in our proposed approach. To solve this problem, our approach needs to consider the order of an opponent's proposals in estimating the opponent's utility. Another important task is to judge the opponent's strategy based on modeling or machine learning technique to further enhance our proposed method.

Reference

1. R. Aydoğan, D. Festen, K.V. Hindriks, C.M. Jonker, Alternating offers protocols for multilateral negotiation, in *Modern Approaches to Agent-based Complex Automated Negotiation*, ed. by K. Fujita, Q. Bai, T. Ito, M. Zhang, R. Hadfi, F. Ren, R. Aydoğan (Springer, To be published)

Agent Buyog: A Negotiation Strategy for Tri-Party Multi Issue Negotiation

Bhargav Sosale, Swarup Satish and Bo An

Abstract The 2015 edition of the Automated Negotiation Agents Competition (ANAC) was the first in its history to introduce multi-party negotiation. To this end, we present the strategy of Agent Buyog, a finalist of the competition. The strategy is based on determining which of the opponent agents is harder to strike a deal with and conceding just enough to please that opponent. This paper aims at outlining various aspects of the strategy such as opponent modeling, concession strategies, bidding strategies and acceptance criteria. It further discusses the limitations of the strategy and discusses possible improvements.

1 Introduction

Negotiation is defined as the procedure of reaching an agreement. Of all modes of conflict management, negotiation has been shown to be the most effective, efficient and flexible. As with many other fields today, there has been an increasing amount of automation in the field by the utilization of intelligent agents. Hence, many research efforts have been made in this regard. The Automated Negotiation Agents Competition (ANAC) [1–4] further fuels such interests.

While previous versions of the competition focused purely on bilateral negotiation, the 2015 edition of the ANAC focused on multiparty negotiation with three

---

B. Sosale · B. An
School of Computer Engineering, Nanyang Technological University, Singapore, Singapore
e-mail: bhargav1@e.ntu.edu.sg

B. An
e-mail: boan@ntu.edu.sg

S. Satish (✉)
Dept. of Electronics and Communication, B.M.S. College of Engineering, Bangalore, India
e-mail: swarupsmail@gmail.com

© Springer International Publishing AG 2017

K. Fujita et al. (eds.), *Modern Approaches to Agent-based Complex Automated Negotiation*, Studies in Computational Intelligence 674,
DOI 10.1007/978-3-319-51563-2_14

agents competing per round. The SOAP protocol [5] was used throughout the competition.

This chapter aims to describe the negotiation strategy used by Agent Buyog, a finalist of the ANAC 2015 competition in both categories; greatest individual utility and greatest social welfare.

2 Agent Buyog Strategy

2.1 Strategy Overview

The agent strategy is modelled on two macro elements; opponent learning, and striking a balance between exploitation and concession.

There is an emphasis on learning the opponent to the best of the agent's ability, based on the ideology that a partially accurate estimate of any aspect of the opponent is better than no estimate at all. By minimizing the unknown it is possible to predict opponent behavior and also propose the most beneficial bid. The aspects of learning include identifying the preference similarities among all parties involved and the concession rate of both opponents. Using the two mentioned aspects the agent is identified with which it is more difficult to strike a deal. One important aspect of the strategy is that both opponents are treated separately at all times. The learning is done separately for each opponent. Any dynamic changes in the behaviour of the opponents are taken into account while proposing our bids.

The second macro element tries to effectively deal with the issue of obtaining the highest utility possible but at the same time keeping negotiation agreement the foremost priority. This is done with the use of a time dependent concession function which decides the lowest acceptable utility for a given moment in time. This effectively creates an expanding window of acceptable bids which expands just enough to reach an agreement with the more difficult agent. The main components of our strategy are as shown in Fig. 1.

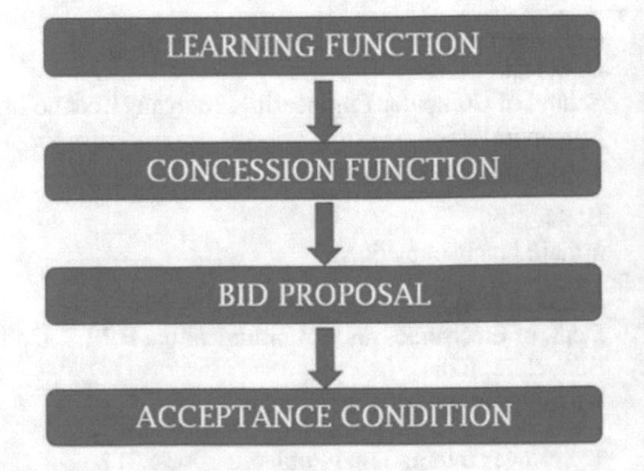

Fig. 1 Strategy overview

2.2 Learning Function

Two learning functions were implemented in Agent Buyog, one for the modeling of preferences and the other for learning the opponents' concession strategy.

The learning of the opponent's concession is based on the strategy described in IAMHaggler [6], where the opponent's concession is modeled in terms of one's own utility. There are two assumptions made; an opponent's utility is inversely proportional to one's own, and the concession curve of an opponent belongs to the family of functions defined by:

$$f(t) = U_0 + e^a t^b$$

Based on the values of a and b, the curve can be used to model both boulware and concessive tactics. $f(t)$ is a function that maps a value of time t to the predicted utility offered by the opponent at that time, with U_0 being the utility of the first bid offered (Fig. 2). The plotting of this curve is performed by statistical regression using weighted least squares, based on the Nelder–Mead Simplex Algorithm. Greater weight is given to newer bids, with the weight of each bid reducing each round. The training data used is a list containing the best bids proposed every round by the opponent. The procedure waits for a predefined number of negotiation rounds before modeling the curve, to avoid inaccuracies resulting from a small sample of training data. Thereafter, the process is repeated every turn.

A simple frequency based model is used to learn the opponent's preferences. It is assumed that the earlier and more frequently certain issue values appear among the opponent's bids, the more the value weighs in its contribution to the opponent's utility.

Each issue is initially assigned an equal weight. The number of unchanged issue values between bids is then measured, and a predefined constant is added to the weights of the unchanged issues. The issue weights are then normalized. The predefined constant is multiplied by the amount of time remaining so that it decreases with

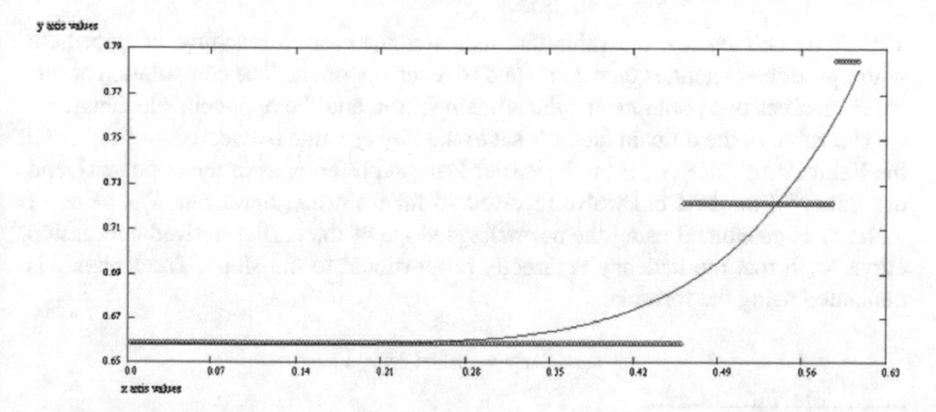

Fig. 2 Curve fitting against a boulware agent

time, ensuring that the earlier appearances translate to higher weights. The decrease of this constant is governed by the equation:

$$\varepsilon_i \times (1 - t^{\alpha+d})$$

In the above equation ε_i refers to the constant value added to each issue weight. α refers to a controlling parameter that describes the nature of the curve and d refers to the discount factor.

For learning the preference over issue values, a similar approach is used. All possible issue values for a particular issue are initialized to 0 and a constant value, ε_v is added to the evaluation of each issue value when they remain unchanged. However, this does not require normalization. They are only normalized to ensure that the evaluation of an issue value does not exceed the maximum permitted evaluation per issue value.

Certain key variables involved in the learning function were parameterized to support experimentation. These values were set after many experimental negotiation sessions. These are:

- Initial estimates of a and b for the non-linear regression. These values were both set to 0.
- The number of iterations of the Nelder–Mead algorithm. This value was set to 5000.
- The number of negotiation rounds to wait before beginning the regression. This value was set to 50.
- The constants to be added to issue weights and issue value evaluations while modeling the preference profile. These values were set to 0.2 and 1, respectively.
- α is set to 1.3.

2.3 Consensus Factor

The *Consensus Factor*, the value that indicates the ease in reaching an agreement with a particular agent, is then computed for each opponent. The computation of this value involves two components; the domain factor, and the opponent's leniency.

The value of the domain factor is set to the higher value between our utility from the Kalai Point[1] (derived from the earlier learned preferences of the opponent), and our utility of the best bid we've received so far from that opponent. The value of leniency is calculated using the normalized slope of the earlier derived concession curve, such that the leniency is directly proportional to the slope. The leniency is computed using the formula:

$$leniency = min(2S_N, 1)$$

[1] Upon empirical testing the learned Kalai Point was found to be roughly 0.1 lesser than the actual Kalai Point. Hence, an offset was made to the learned Kalai Point.

Here, S_N refers to the normalized slope of the learned concession curve.
Using the two components, the Consensus Factor is then computed as:

$$Consensus\ Factor = w_l \times leniency + w_d \times domain\ factor$$

Here, w_l and w_d are the weights given to each component and are decided based on the two formulae:

$$w_d = 1 - leniency^{\gamma}$$

$$w_l = 1 - w_d$$

The parameter γ represents a value that defines how the weight assigned to the best agreeable bid changes with respect to the leniency. This value is set to 1.75.

The weights assigned are dynamic, so as to ensure that our agent neither under-exploits nor gets exploited by the opponent agent. Assigning weights also helps offset errors that may occur from our initial assumption that the opponent's utility is inversely proportional to our own.

The Consensus Factor is calculated for both opponents separately at every round of the negotiation, ensuring that our agent responds immediately to any change in opponent behaviour.

2.4 Concession Curve

The concession curve (Fig. 3) is derived using the lower of the two Consensus Factors using the formula:

$$f(t) = Consensus\ Factor + (1 - Consensus\ Factor) \times (1 - t^{\beta})$$

This allows our agent to concede just enough to satisfy the opponent with the lower agreeability.

The concession curve follows a standard time dependent concession as explained in [7]. This curve remains flexible, being recalculated at every round of the negotiation process, due to the dynamic nature of the Consensus Factor. This allows our agent to respond to changes in opponent behaviour immediately. To allow for discounted domains, the agreeability value is multiplied by the discount factor. This ensures faster concession in the case of a discounted domain.

The optimal value for β that influences the nature of the curve was determined by empirical evaluation to be 1.8.

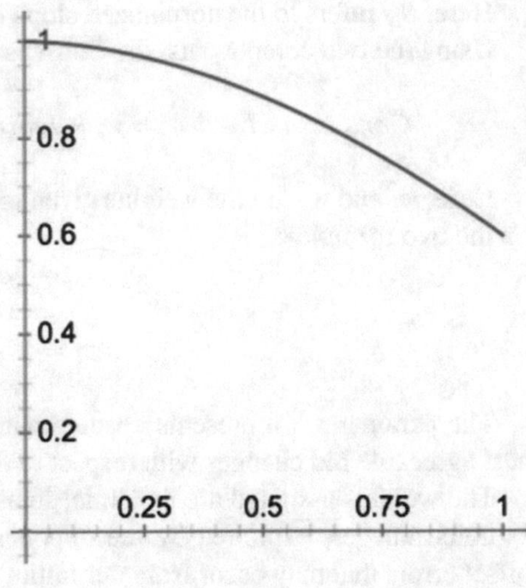

Fig. 3 Time concession curve

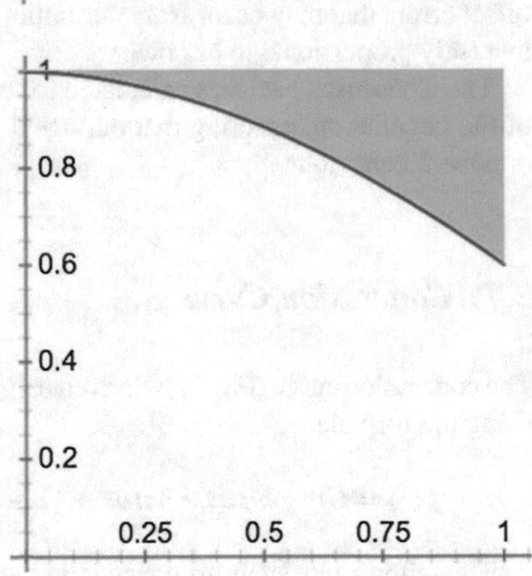

Fig. 4 Bidding strategy

2.5 Bidding

The bidding strategy used by Agent Buyog is based on a system that uses an expanding window of bids. This expanding window, as shown in the Fig. 4, is derived from the agent's time-dependent concession curve. The lower end of the bidding window is set to $f(t)$ from the time-dependent concession curve, while the upper end of the bidding window is always set to 1. This gives the agent an increasing range of bids to choose from at every round of the negotiation and the agent then aims to determine the most optimal bid to propose. For each bid in the range, the agent then:

1. Calculates the utility value the bid provides to each opponent, using the previously learned preference profiles.
2. Calculates the weighted Euclidean distance between the utilities provided to the opponent and the point (1, 1) in the opponents' utility space.
3. The bid least farthest to (1, 1) in the opponents' utility space is then proposed.

The weights provided in the calculation of the Euclidean distance favour the more difficult agent by using the formula:

$$\sqrt{(1 - Consensus\ Factor_A)(U_A - Kalai_A)^2 + (1 - Consensus\ Factor_B)(U_B - Kalai_B)^2}$$

Here U_A, U_B refer to utilities gained by opponents A and B for an arbitrary bid.

To increase chances of agreement, as well as save computational time, the agent first checks any bids in the window previously agreed upon by both opponents before checking the rest of the window.

2.6 Acceptance

Agent Buyog follows a simple acceptance mechanism. The criteria for accepting a bid proposed by an opponent is as follows:

- The bid utility must lie in the acceptance window.
- It must not yield our agent a utility lesser than any bid previously agreed upon by both opponents.
- The utility of the bid must be greater than or equal to the subsequent bid due to be proposed by our agent.

2.7 Miscellaneous

2.7.1 Walk Away Strategy

The negotiation is terminated when the expanding window reaches the undiscounted reservation value. At this point it is ascertained that the negotiation has reached its saturation point and no beneficial agreement can be reached. Terminating the negotiation is largely useful in scenarios where both opponents are unable to reach an agreement with each other, despite our agent reaching a separate consensus with both.

2.7.2 Desperate Concession

When the negotiation reaches its final moments (number of rounds remaining <3) the bottom end of the acceptance window is reduced by 50%. This concession factor is derived empirically. This ensures that in the dying moments of the negotiation, agreement is prioritized over exploiting the opponents.

3 Conclusion and Future Works

An effective negotiation agent for use in tri-party negotiations based on the SOAP protocol was introduced. The agent described, works on the basis of an adaptive concession curve aiming to strike a balance between exploitation and concession based on the behaviour of the opponents. Learning the concession behaviours and the preferences of both opponents, a Consensus Factor is computed and an optimal bid is ascertained. This maximizes the probability that the bid will be accepted by both opponents.

Despite displaying reasonable results in the competition there are certain aspects of the strategy that require further research.

The Nelder–Mead iterative algorithm greatly depends on the initial estimates provided. Worse initial estimates resulted in extremely inaccurate predictions. It also has the tendency to get stuck in local optima. These drawbacks could make the algorithm largely unreliable, and better approaches to curve fitting could be tested. Alternative regression methods such as Gaussian Regression could be tested with the agent to overcome these drawbacks.

Our strategy is predominantly based on treating the two opponents as separate entities during the process of learning. An alternate strategy could involve treatment of both opponents as a single entity, working with a representation of the average behaviour exhibited by both opponents. This could greatly reduce the computational time involved in learning.

Other worthwhile directions could include more systematic and detailed testing of the various parametric values used. These parameters could greatly affect the agent's performance.

References

1. T. Baarslag, K. Fujita, E.H. Gerding, K. Hindriks, T. Ito, N.R. Jennings, C. Jonker, S. Kraus, R. Lin, V. Robu, C.R. Williams, Evaluating practical negotiating agents: results and analysis of the 2011 international competition. Artif. Intell. **198**, 73–103 (2013)
2. T. Baarslag, K. Hindriks, C.M. Jonker, S. Kraus, R. Lin, The first automated negotiating agents competition (ANAC 2010), in *Studies in Computational Intelligence* (2012), pp. 113–135
3. K. Gal, L. Ilany, The fourth automated negotiation competition, in *Next Frontier in Agent-based Complex Automated Negotiation* (2015), pp. 129–136

4. C.R. Williams, V. Robu, E.H. Gerding, N.R. Jennings, An overview of the results and insights from the third automated negotiating agents competition (ANAC2012), in *Novel Insights in Agent-based Complex Automated Negotiation* (2014)
5. R. Aydogan, D. Festen, K.V. Hindriks, C.M. Jonker, Alternating offers protocols for multilateral negotiation, in *Modern Approaches to Agent-based Complex Automated Negotiation*
6. C.R. Williams, V. Robu, E.H. Gerding, N.R. Jennings, IAMhaggler: a negotiation agent for complex environments. New Trends Agent-Based Complex Autom. Negot. **383**, 151–158 (2012)
7. P. Faratin, C. Sierra, N.R. Jennings, Negotiation decision functions for autonomous agents. Robot. Auton. Syst. **24**(3–4), 159–182 (1998)

Phoenix: A Threshold Function Based Negotiation Strategy Using Gaussian Process Regression and Distance-Based Pareto Frontier Approximation

Max W.Y. Lam and Ho-fung Leung

Abstract Automated negotiation is of great interest in artificial intelligence. An effective automated negotiation strategy can facilitate human in reaching better negotiation outcomes benefiting from the adoption of advanced computational methods. This paper deals with multi-lateral multi-issue negotiation where opponents' preferences and strategies are unknown. A novel negotiation strategy called *Phoenix* is proposed following the negotiation setting adopted in *The Sixth International Automated Negotiating Agents Competition* (ANAC 2015) [13]. In attempt to maximize individual utility and social welfare, we propose two highlighted methods – *Gaussian Process Regression* (GPR) and *Distance-based Pareto Frontier Approximation* (DPFA). Integrating the idea of these methods into a single function called threshold function, we show that *Phoenix* is a fully adaptive, cooperative and rationally designed strategy.

1 Introduction

Negotiation is naturally a humanistic solution to conflicts between different parties. It has long been studied in game theory and economics [12]. Owing to many potential applications in industrial and commercial domains, automated negotiation becomes a blossoming research area in artificial intelligence. Benefiting from the adoption of advanced computational methods, automated negotiation techniques can significantly alleviate human efforts, and are capable of tackling complex negotiations which is difficult to be resolved by human beings.

Until now, a number of strategies have been proposed. Yet, many works are dealing with bilateral negotiation [1, 2, 7], where, for instance, the consumer-provider or

M.W.Y. Lam (✉) · H.-f. Leung
Department of Computer Science and Engineering,
The Chinese University of Hong Kong, Shatin, Hong Kong, China
e-mail: wylam3@cse.cuhk.edu.hk

H.-f. Leung
e-mail: lhf@cse.cuhk.edu.hk

© Springer International Publishing AG 2017

K. Fujita et al. (eds.), *Modern Approaches to Agent-based Complex Automated Negotiation*, Studies in Computational Intelligence 674,
DOI 10.1007/978-3-319-51563-2_15

buyer-seller relationship is concerned [9]. On the contrary, multilateral negotiation where more than two agents negotiate against each other has received far less attention, despite its generality. This is very much due to the complexity of analytically evaluating a broad spectrum of negotiation strategies under multilateral negotiation. However, one indeed can rely on simulations between different strategies. To this end, the *International Automated Negotiating Agents Competition* (ANAC) was organized annually starting from 2010 [13] providing a benchmark for evaluating various negotiation strategies. In ANAC, negotiation is simulated under a well-developed platform, GENIUS [8], where realistic negotiation setting is adopted and agents' performances are extensively evaluated in assorted multi-issue negotiation.

In this paper, we are interested in multi-lateral multi-issue negotiation. A novel negotiation strategy called *Phoenix* is proposed. The design of *Phoenix* follows the negotiation setting adopted in *The Sixth International Automated Negotiating Agents Competition* (ANAC 2015) [13]. There are two objectives in this competition – maximizing individual utility, and enhancing social welfare (i.e. the largest sum of scores achieved by two strategies). In the remainder of this paper, we will show how *Phoenix* achieve these two goals by using the two highlighted methods – *Gaussian Process Regression* (GPR) [10] and *Distance-based Pareto Frontier Approximation* (DPFA). To facilitate the work of analysis and to make our strategy expressible, we incorporate these two methods into a single continuous function, namely, threshold function. As we will show, all informative factors that can be obtained in the negotiation are considered to construct the threshold function optimally.

The remainder of this paper is organized as follows. In Sect. 2, we overview some related work to *Phoenix*. In Sect. 3, we describe the formulation of *Phoenix* in details. In Sect. 4, we analyze the performance of *Phoenix* with experimental settings. Finally, in Sect. 5, we conclude our work and identify potential future research directions.

2 Related Work

Recent research on developing negotiation strategies for automated negotiating agents is of growing interest in the negotiation community. While the ANAC competition successfully encourages a variety of designs of practical negotiation agents, Baarslag et al. [2] generalize these negotiation strategies with the proposition of *BOA* framework. As it literally means, *BOA* consists of three functional components – Bidding Strategy, Opponent Model and Acceptance Strategy. With generic framework, works can be simplified into three independent researches, while each exists many approaches in the literature.

Although *Phoenix* is not structured following the *BOA* framework, its success is very much due to the bidding strategy DPFA and the opponent model GPR. To our knowledge, this paper is a pioneer work to formulate the DPFA mechanism. Yet, GPR indeed gains much attention as an state-of-the-art modeling technique [10]. Related works to this opponent-modelling technique are described below.

Back to earlier research on bargaining problem, Rubinstein [12] suggests that modelling opponents is the key importance to achieve good performance. Such belief remains till now. A board spectrum of opponent-modelling techniques is applied in automated negotiation. Thielscher and Zhang [1] summarize those used in bilateral automated negotiation, showing that many of them are indeed borrowed from machine learning. In attempt to generalize the design of opponent models, Hindriks and Tykhonov [7] propose a generic framework using Bayesian Learning.

Regarding our proposed modelling technique, GPR [10] is also a Bayesian Learning model being renowned in the field of machine learning as a principled method to model functions by performing non-linear non-parametric regression. In particular, GPR is an extremely useful technique to model trends from time-series data [3, 5, 11], which exactly fits the task of opponent modelling. In the light of this, Williams et al. [14] formulate an opponent-modelling method making use of GPR. What's more, many ANAC participants [4, 15, 16] also employ GPR attaining excellent results in the competition.

3 The *Phoenix* Strategy

In this section, we will explore our proposed strategy *Phoenix*, which is composed of two functional components. The first component is the *Threshold Function Construction* (TFC) component. It is responsible for determining our threshold function $f_{\text{thre}}(t)$, which denotes the change of our agent's attitude throughout the negotiation, and is defined continuously by the time variable $t \in [0, 1]$. To be concise, the main usage of the threshold function is to represent a bottom line of utilities of bids that we can accept or offer. Being of utmost importance in *Phoenix*, we harness all informative factors that can be obtained in the negotiation, including opponent model using GPR, to build this function optimally. Details are shown in Sect. 3.1. The second component is the *Decision Making* (DM) component, which is responsible for the decision of next move in each round. As mentioned, this decision process, to a great extent, relies on the threshold function that we constructed in the first component. A noteworthy point in this component is the use of DPFA in bids proposal. This method is proposed to address the incentive of enhancing social welfare. The scheme is detailed in Sect. 3.2.

3.1 Threshold Function Construction (TFC) Component

In *Threshold Function Construction* (TFC) component, our goal is to construct an appropriate threshold function for later use. As we briefly discussed, $f_{\text{thre}}(t)$ indeed implies agent's attitude over time. Therefore, concession is nothing but a gradually decreasing threshold function. To formulate rational concession-making behaviour, we define the concession rate $r(t)$ at time t, and the lowest concession utility u_{low}, that is, the lowest value that our threshold function can reach.

The question now resorts to what we should consider for making concession. In fact, it really depends on what information our agent can obtained from the negotiation environment. In *Phoenix*, we focus on 4 consideration factors:

1. **Discounting factor** δ. The effect of discounting factor on the final utility is given by $u_{\text{ag}} \cdot \delta^{t_{\text{ag}}}$, where u_{ag} is the undiscounted utility of the offer that all parties agreed, and t_{ag} is the agreement time. Therefore, we should reach an agreement as soon as possible. In order words, when the discounting factor is small, we should increase $r(t)$ and decrease u_{low}.
2. **Reservation value** θ. The reservation value is the utility obtained when the negotiation is terminated, or the deadline is reached. In such cases, we will gain $\theta \cdot \delta^{t_b}$, where t_b is the time of breaking negotiation. From this fact, small reservation value is a threat when the deadline is about to be reached. Therefore, intuitively, we should increase $r(t)$ when the t is getting close to 1.
3. **The distribution of bids' utilities**. The distribution of bids' utilities in fact is the most reasonable guideline to justify how good a received offer is. For example, an offer with utility greater than average should be considered as reasonably good. Thus, this is an important argument of u_{low}. For later use, we denote $\mu_{\mathscr{B}}$ as the corresponding mean, and $\sigma_{\mathscr{B}}$ as the corresponding standard deviation.
4. **Opponents' attitudes**. Since we concern about mutual benefits, it is preferable to make our decisions in accordance with opponents' attitudes during the negotiation process. Yet, opponents' attitudes are somewhat abstract in this sense. Thus, we measure opponents' attitudes using the trend of opponents' concession. Then, we can adaptively make concessions following this trend.

In fact, the calculations of the first three factors are trivial given the negotiation domain. However, the calculation of the forth factor is more difficult due to the need for opponent modelling. Regarding this task, GPR is employed. One should be careful that our formulation of modelling task is different from the previous works of using GPR [4, 15, 16], since our objective is not to model opponents' preferences but to model the trend of their concession from our observation. For better reasoning, we first have to make two fundamental assumptions about opponents:

Assumption 1 Opponents propose offers following some predefined function.

Assumption 2 Opponents' offers are uniformly selected from the bids which have utilities greater than their predefined function.

With these assumptions, it is now reasonable to employ GPR to model the trend of concession as a composition of opponents' predefined functions.

To train the GPR, we need to collect some data to build training data set. In our case, the concession is of our main concern. Intuitively, this is computed by analyzing the offers proposed by all opponents. Say P opponents proposing P offers each round, we only pick the one with the minimum utility to be a conservative estimator of opponents' concession. This value indeed holds an important meaning.

If the utility of our offer is below it, our opponents will be very likely to accept our offer. It is very handful for us to control when an agreement should be made, such that our gain is optimized. Formally, we define this estimator as a $\hat{c}(t)$ where t is the time of computation. Note that the estimator collected each round is not yet suitable to be the training data, since as mentioned in Assumption 2 opponent's offer is proposed with uniformly distributed noise. Fortunately, it is possible to denoise the data with the following proposition:

Proposition 1 *Among a set of computed estimator $\hat{c}(t)$, the maximum one is of the highest probability to be the closest one to the opponents' least threshold.*

We divide the time interval of whole negotiation into K equal sub-intervals and denote the boundaries by $K+1$ timestamps $\{t_0, t_1, t_2, \ldots, t_K\}$, where

$$0 = t_0 < t_1 < t_2 < \cdots < t_K = 1. \tag{1}$$

Since only opponents' least threshold can be used as a information of their attitudes, we use Proposition 1 to obtain the denoised estimator for concession in k-th time sub-interval. That is,

$$y(k) = \max_{t_{k-1} < t \leq t_k} \hat{c}(t). \tag{2}$$

When the negotiation reaches the k-th time sub-interval, we obtain a sequence of selected scores, which yields our training data set for GPR

$$\mathscr{T}(k) = \left\{ \left(\frac{t_0 + t_1}{2}, y(1) \right), \left(\frac{t_1 + t_2}{2}, y(2) \right), \ldots, \left(\frac{t_{k-1} + t_k}{2}, y(k) \right) \right\}. \tag{3}$$

After the training of GPR, we will obtain a mean function denoting the expected trend, and the 95% confidence interval around the mean function, as showed in Fig. 1. For later use, we denote $f_{\mathrm{gp}}(t)$ as the mean function.

Now, we finally obtain all information needed to construct the threshold function as discussed previously. Firstly, we derive the concession rate

$$r(t) = \left(t^{\delta} + (1 - \theta) t^4 - \alpha \frac{d}{dt} f_{\mathrm{gp}}(t) \right), \tag{4}$$

where $\alpha \in [0, 1]$ is defined as the *degree of sensitivity*. The higher α we set, the more sensitive to opponents' offers. Secondly, we compute the lowest concession

$$u_{\mathrm{low}} = \max\left(\theta, \mu_{\mathscr{B}} - 2(1.5 - \delta)\sigma_{\mathscr{B}}\right). \tag{5}$$

Note that the setup of u_{low} can prevent any irrational concessions induced by GPR, as we may be misled by the offers from opponents, thus u_{low} do not depend on $f_{\mathrm{gp}}(t)$.

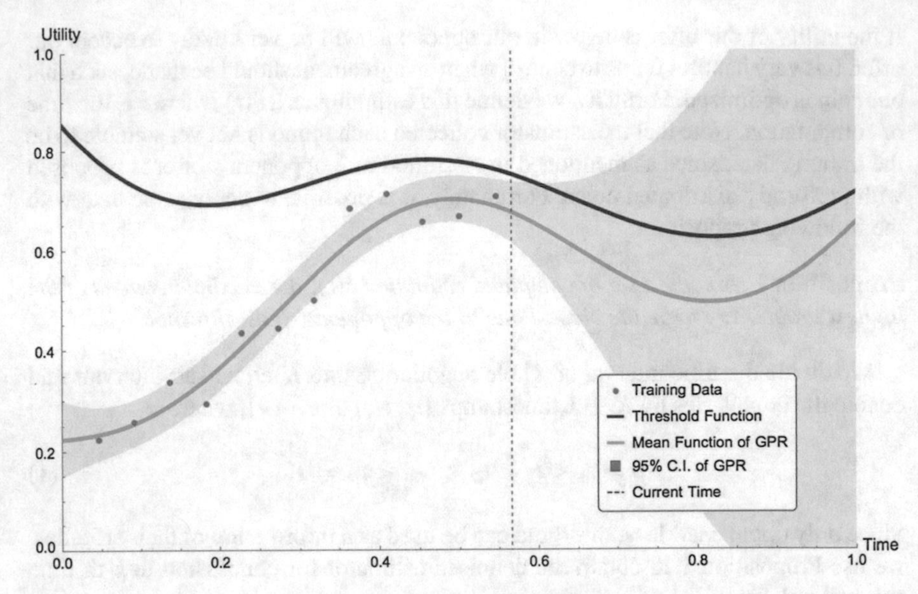

Fig. 1 An example of constructing threshold function using the results from GPR and variables from the domain

Finally, we formulate our threshold function

$$
\begin{aligned}
f_{\text{thre}}(t) &= u_{\text{low}} + (1 - u_{\text{low}}) \left[\max\left(0, \int -r(t)dt\right) \right] \\
&= u_{\text{low}} + (1 - u_{\text{low}}) \left[\max\left(0,\ \alpha f_{\text{gp}}(t) - \frac{t^{1+\delta}}{1+\delta} - \frac{(1-\theta)t^5}{5} + \beta\right) \right],
\end{aligned}
\tag{6}
$$

where $\beta \in [0, 1]$ is the *degree of exploitation* controlling how conservative our agent will be. Note that the threshold function is updated whenever we shift to a new time sub-interval and get a new training data set for GPR.

An example showing how the output of GPR affecting threshold function is shown in Fig. 1. Notice that the threshold function indeed always resorts to a U-shape curve because opponents tend to concede more when the negotiation deadline is approaching. Interestingly, this U-shape curve matches the strategy proposed by Hao et al. [6] called ABiNes, which won ANAC 2012. As our formulation of threshold function using GPR gives a logical explanation of their success, we conceive that *Phoenix* is a generic strategy.

3.2 Decision Making (DM) Component

Decision Making (DM) component is the second component of *Phoenix*, which is responsible for choosing an optimal move to proceed the negotiation. Normally, there are in total three possible choices – (1) acceptance, (2) termination, and (3) new offer proposal.

We divide the task of considering these three choices into three standalone functions. In particular, the first two are Boolean functions. It checks whether the criteria of executing such move have been met, where the criteria of acceptance and termination are discussed in Sects. 3.2.1 and 3.2.2 respectively. Note that only when these two conditions are not met, we proceed to the third function, which rationally searches bids to propose. The searching algorithm is explained in Sect. 3.2.3.

3.2.1 Acceptance Function

The mission of acceptance function is to decide whether our agent should accept the received offer or not. In fact, the checking is very simple once we have determined the threshold function, since the threshold function is exactly decision boundary of acceptance. Therefore, if the utility of our received offer is greater than the value of threshold function $f_{\text{thre}}(t_c)$ at time t_c, such an offer will be accepted.

3.2.2 Termination Function

Termination Function is obligated to determine whether our agent should terminate the negotiation or not. To begin with, we first investigate the impact of termination. As discussed previously, once our agent determined to terminate, the negotiation will end immediately and each party will be given the final utility $\theta * \delta^{t_b}$. As a result, we terminate the negotiation if or only if there exist $t \in [t_c, t_c + \tau]$ such that $f_{\text{thre}}(t) \leq \theta$. Here, we consider not only the current time t_c but also a period of time further using the *forewarning contant* τ. It implies that later received offers is going to be worse in the future, therefore the negotiation should soon be terminated.

3.2.3 Offer-Proposing Function

Offer-proposing Function is called when the above Boolean functions both are of false results. In this case, we have to propose a new offer to our opponents such that the chance of our offer being accepted is maximized, that is, a mutually beneficial offer.

In Economics, such offers are defined as *Pareto frontiers*.[1] Indeed, with no prior information about opponents' preferences, it is impossible to analytically find the

[1]Pareto frontiers is a set of Pareto optimal bids in terms of utility.

exact Pareto frontiers. However, approximation methods do exist. In our work, an approximation algorithm called DPFA is used to address this issue. The overall flow of DPFA is shown on Algorithm 1. By looking at opponents' offering patterns, rating for each bid is calculated such that higher rating is more appealing to our opponents. Details of DPFA are described below.

Algorithm 1 DPFA – an approximation algorithm to search for Pareto frontiers.

Input: Negotiation domain $\mathscr{D}$, Threshold function $f_{\text{thre}}(t)$
Output: Bid to propose $\mathbf{b}^*$

```
1: procedure SEARCHBESTOFFER
2:     𝒝 ← getAvaliableBids(𝒟, f_thre(t));
3:     ℛ ← getReferenceBids(𝒟);
4:     for each bid b_j in 𝒝 do
5:         x_j ←computeRating(b_j, ℛ);
6:     x ← (x_1, x_2, ..., x_|𝒝|);
7:     b* ← drawBidFollowRating(𝒝, x);
```

Firstly, only the bids with utilities greater than $f_{\text{thre}}(t)$ will be considered below. Such bids are defined as *available bids*, $\mathscr{B}$. Besides, we assume that there are some bids very probable to be one of the Pareto frontiers and can be easily obtained. Such bids are defined as *reference bids*, $\mathscr{R}$. In our work, reference bids are selected from: (1) opponents' first offers, (2) opponents' best offers ever, and (3) opponent's previous offer. In total, with P opponents we have $3P$ reference bids.

Regarding (1), in practice, our opponent's first offer is very likely to be his best offer, which is one of Pareto frontiers. What's more, (2) is also sensible. When our opponents propose bids following some threshold, the best ones that we receive should be the close to the Pareto frontiers. In addition, we assume that our opponents also search for Pareto frontiers. Then, (3) is also an estimator of the Pareto frontier, and this should become more accurate as negotiation processes. However, noticing that our opponents propose bids with some randomness, (3) should not be weighted equally as the other two.

Note that in the negotiation setting of ANAC [8], each bid is represented as a vector of issue values. Therefore, we can approximate how close each bid $\mathbf{b}_j \in \mathscr{B}$ to Pareto frontiers by measuring its Euclidean distance to all reference bids $\mathbf{r}_i \in \mathscr{R}$. This yields

$$\text{rating}(\mathbf{b}_j) = -\sum_{i=1}^{3P} \gamma_i \|\omega \circ (\mathbf{b} - \mathbf{r}_i)\|_2 \ , \tag{7}$$

where $\circ$ is the Hadamard product (element-wise product), $\gamma_i \in [0, 1]$ is the predefined *salience* for the i-th reference bid, and ω is the issues-weighting vector to underline the importance of each issue before calculating the Euclidean distance, as we know that the issues are not equally weighted. However, as one should expect, ω is very difficult to be derived analytically. Fortunately, showed with success in [4, 6], the frequency-based method works effectively in issue-weighting problem. In *Phoenix*, we initially set ω as a vector of ones. Then, in each round of the negotiation, we update the i-th entry ω_i by summing the number of received offers having equivalent value in the i-th issue.

After the rating scores are calculated, we do not directly choose the bid with highest rating score as our next offer. Instead, we choose it randomly following some probability distribution, where bids with higher rating have higher probability to be chosen. With this fuzzy logic, we can avoid proposing bids repeatedly; while in practice some bids may have particularly high rating score due to stochasticity.

4 Performance Analysis

In order to evaluate *Phoenix*, we have implemented a negotiation agent called *PhoenixParty* to compete in the ANAC competition [13]. While *PhoenixParty* is the first work applying the *Phoenix* strategy in automated negotiation, we experimentally hand tuned the parameters with our intuition described in Sect. 4.1. Following this, in Sect. 4.2, the results corresponding to this parameters setting are analyzed.

4.1 Parameters Setting

In total, there are 4 essential parameters in *Phoenix*: degree of sensitivity α (Eq. 4), degree of exploitation β (Eq. 6), forewarning constant τ (Sect. 3.2.2), and salience of reference bids γ (Eq. 7). Interestingly, these parameters are sufficient to determine what type of agent we are, i.e., *Conceder* or *Boulware*. As a trial, in *PhoenixParty*, we set $\alpha = 0.3$, $\beta = 0.9$, $\tau = 0.1$ and $\gamma = [1\ \ 0.8\ \ 0.3]^P$, where the i-th entry in γ correspond to $3P$ reference bids introduced in Sect. 3.2.3.

4.2 Results and Analysis

Following the rules of ANAC2015 [13], the evaluation process on participating agent is separated into two phases. The first phase is the qualifying round, where 24 participants are assigned to four pools randomly. In each pool, 6 agents are competing against each other such that only the two agent with highest individual utilities are able to proceed to the final round. As shown in Table 1, *PhoenixParty* was assigned to Pool 4, and achieved second highest individual utility in the pool.

Regarding the results in qualifying round, there actually is a thought-provoking phenomenon – the resulted utility in Pool 4 is much lower than that in other 3 pools. Note that the agents in all pools negotiate under identical negotiation domains with equivalent bids' utilities. Therefore, the trigger of resulting lower utility in Pool 4 should be late agreements in discounted domains. Surprisingly, this fact matches our

Table 1 Agents' average individual utilities in qualifying round [13]

Agent name	Average utility	Agent name	Average utility
Pool 1		Pool 2	
AgentBuyogV2	0.597955067	Atlas3	0.680664517
PokerFace	0.594266467	XianFaAgent	0.633863800
PNegotiator	0.591739600	MeanBot	0.584072250
DrageKnight	0.571351533	AgentX	0.571492517
Mercury	0.550937867	AgentHP	0.535089883
SENGOKU	0.547276433	TUDMixed	0.504430117
Pool 3		Pool 4	
ParsAgent	0.582228250	RandomDance	0.408558450
Kawaii	0.575404450	**PhoenixParty**	**0.380885900**
Group2	0.567122400	AresParty	0.378801767
CUHKAgent2015	0.552638067	AgentNeo	0.356815667
AgentW	0.518159433	AgentH	0.339627333
JonnyBlack	0.491797117	Forseti	0.258990217

parameter setting of *PhoenixParty*, as $\beta \in [0, 1]$ is set to be 0.9. Theoretically, this should be the direct consequence of high degree of exploitation, since our agent tends to exploit opponents and seldom makes concession. If this is the case, our agent will delay the agreement time and thus the negotiating parties in the same pool are likely to result very low individual utilities in discounted domains.

Indeed, the result of final round also gives evidences to our agent's exploitation behaviour. As shown in Table 2, the individual utilities of the agents from Pool 1, Pool 2 and Pool 3 are significantly deceased compared to their performances in qualifying round. On the contrary, the performances of agents from Pool 4 – *RandomDance* and *PhoenixParty* are significant improved in the final round. Again, this interesting phenomenon can be explained by our agent's exploitation behaviour. Without surprise, agents without any exploitation behaviour are probable to reach early agreement in discounted domain. Therefore, albeit our agent can exploit them successfully, our individual utilities can still be very low after averaging.

Overall, the results of ANAC2015 show that setting high degree of exploitation can result in severe exploitation behaviour, which is prone to delay the agreements because of persistence in getting our favourable bids. In this case, despite our agent can get better offers than opponents in each negotiation session, after considering the late agreements from discounted domains, the average resulting utility may become worse. From the experimental parameter settings in *PhoenixParty*, we expect the degree of exploitation to be lowered in discounted domain. In fact, it is more proper if we carefully tune its value by running more testing, noticing the crucial influence of this parameter on the resulting utilities.

Table 2 Agents' average individual utilities in final round with comparison to their performances in Qualifying Round [13]

Final round

Agent name	Average utility	Change from qualifying round
Atlas3	0.481042722	(−29.3274865%)
ParsAgent	0.470693979	(−19.1564517%)
RandomDance	0.46062548	(+12.7440834%)
Kawaii	0.460129481	(−20.0337294%)
AgentBuyogV2	0.458823101	(−23.2620426%)
PhoenixParty	**0.442975836**	**(+16.3010073%)**
XianFaAgent	0.353133027	(−44.2024544%)
PokerFace	0.344003999	(−42.1128369%)

5 Conclusions and Future Work

In this paper we describe a negotiation strategy called *Phoenix* for multi-lateral automated negotiation. The entire strategy is based on a time-dependent continuous function, namely, the threshold function. This function denotes our agent's attitude and is referred whenever our agent needs decides the next move. It cannot only simplify the decision process, but also make agent's decisions analytically tractable by plotting graphs. To derive an appropriate threshold function, *Phoenix* harnesses all possible information perceived in the negotiating environment as inputs to threshold function. Being adaptive to various opponents' strategies, opponents' behaviours are also considered using opponent-modeling technique.

Through success, our approach is greatly highlighted and detailed in this paper. A well-designed modelling task is formulated taking opponents' behaviours into account. With our formulation, time-series data are collected throughout the negotiation. We borrow a sophisticated modelling technique specializing in time-series prediction called *Gaussian Process Regression* (GPR). Note that GPR is a flexible modelling framework which fully specified by our prior setting. In fact, the output of GPR is a distribution over function, where the functional properties are dominated by the kernel function in prior setting. Therefore, the performance of various kernel functions applying on our formulated modelling task still need to be investigated. Also, since the prediction of GPR is not guaranteed to capture opponents' future attitude, analysis is required to determine its correctness. Utilizing the output of GPR, all considerations putting on opponents are addressed. Together with other considerations putting on negotiation domains, an all-round threshold function can eventually be derived.

Once a proper threshold function is derived, the work remaining will be deciding agent's next move. Indeed, using the threshold function, the determination of acceptance and termination is somewhat trivial. In fact, the noteworthy part of deciding agent's next move is how *Phoenix* propose the next offer for opponents. With the incentive for maximizing mutual benefits, the second highlight of our work – the *Distance-based Pareto Frontier Approximation* (DPFA) method is introduced in this paper. To our knowledge, this paper is a pioneer work to formulate the DPFA mechanism, which uses distance in issue space to approximate the Pareto Frontier. In practice, this method can achieve reasonably accurate results, though it has not guarantee to be true in all cases. Indeed, the main difficulty of employing DPFA is setting the parameters without loss of generality. This remains an active area for future research.

References

1. T. Baarslag, M. Hendrikx, K. Hindriks, C. Jonker, Measuring the Performance of Online Opponent Models in Automated Bilateral Negotiation (2012)
2. T. Baarslag, K. Hindriks, M. Hendrikx, A. Dirkzwager, C.M. Jonker, Decoupling negotiating agents to explore the space of negotiation strategies, in *Novel Insights in Agent-based Complex Automated Negotiation* (2014)
3. S. Brahim-Belhouari, A. Bermak, Gaussian Process for Nonstationary Time Series Prediction (2004)
4. S. Chen, H.B. Ammar, K. Tuyls, G. Weiss, Optimizing Complex Automated Negotiation Using Sparse Pseudo-Input Gaussian Processes (2013)
5. J. Cunningham, Z. Ghahramani, C.E. Rasmussen, Gaussian Processes for Time-Marked Time-Series Data (2012)
6. J. Hao, H.F. Leung, Abines: An Adaptive Bilateral Negotiating Strategy Over Multiple Items (2012)
7. K. Hindriks, D. Tykhonov, Opponent Modelling in Automated Multi-issue Negotiation Using Bayesian Learning (2008)
8. R. Lin, S. Kraus, T. Baarslag, D. Tykhonov, K. Hindriks, C.M. Jonker, Genius: An Integrated Environment for Supporting the Design of Generic Automated Negotiators (2014)
9. F. Lopes, M. Wooldridge, A.Q. Novais, Negotiation Among Autonomous Computational Agents: Principles, Analysis and Challenges (2008)
10. C.E. Rasmussen, C.K. Williams, *Gaussian Processes for Machine Learning* (The MIT Press, Cambridge, 2006)
11. S. Roberts, M. Osborne, M. Ebden, S. Reece, N. Gibson, S. Aigrain, Gaussian Processes for Time-Series Modelling (2013)
12. A. Rubinstein, Perfect Equilibrium in a Bargaining Model (1982)
13. The Sixth International Automated Negotiating Agents Competition (ANAC 2015) (2015), http://www.tuat.ac.jp/~katfuji/ANAC2015/
14. C.R. Williams, V. Robu, E.H. Gerding, N.R. Jennings, Using Gaussian Processes to Optimise Concession in Complex Negotiations Against Unknown Opponents (2011)
15. C.R. Williams, V. Robu, E.H. Gerding, N.R. Jennings, Iamhaggler: A Negotiation Agent for Complex Environments (2012)
16. C.R. Williams, V. Robu, E.H. Gerding, N.R. Jennings, Iamhaggler2011: A Gaussian Process Regression Based Negotiation Agent (2013)

Pokerface: The Pokerface Strategy for Multiparty Negotiation

J.B. Peperkamp and V.J. Smit

Abstract In this paper we will discuss a possible strategy to use in multiparty negotiation. We show why this scenario differs from bilateral negotiation and describe our strategy that deals with these differences, its merits and shortcomings and possible future improvements.

1 Introduction

This paper describes the negotiation strategy of the agent Pokerface which took part in the finals of the Sixth International Automated Negotiating Agents Competition (ANAC2015). In this fifth edition of the competition, the challenge was to develop strategies that can be used in negotiations with more than one opponent. Clearly it is not possible to apply simple strategies for two-party negotiation to this multiparty scenario without any modification. Take for instance the tit-for-tat approach [3]: if two opponents act in opposite ways, an agent following this strategy would not know which one to mimick. A new protocol is also used to facilitate the multiparty negotiation; see [1]. As in previous editions (see e.g. [2, 4, 5, 7]), the strategy is written as an agent that performs its negotiation via the Genius framework [6].

2 Strategy

A brief overview of the Pokerface strategy that the eponymous agent uses will now be given. The basic idea is to throw off opponents by performing seemingly random actions for the first part of the negotiation, hence the name of the strategy. After this

J.B. Peperkamp (✉) · V.J. Smit
Delft University of Technology, Delft, The Netherlands
e-mail: jbpeperkamp@gmail.com

V.J. Smit
e-mail: vjsmit@gmail.com

© Springer International Publishing AG 2017

K. Fujita et al. (eds.), *Modern Approaches to Agent-based Complex Automated Negotiation*, Studies in Computational Intelligence 674,
DOI 10.1007/978-3-319-51563-2_16

a straightforward polynomial concession strategy is employed, with a twist at the end. An opponent model is also specified, although this was not the main focus of the work.

2.1 First Stage: Random Walk

As mentioned, the first part of the negotiation is spent offering ostensibly random bids to throw off analysis that opponents may be trying to perform on the agent's actions. Of course the bids cannot be entirely random, because then a large concession might happen at some point that leads to an undesirable outcome for the agent. To prevent this from happening, the agent does take into account its utilities for each issue value when constructing the random bids. Only when the total utility of the issues in the randomly selected bid is above a predefined threshold of desirability will the bid be used, otherwise it is excluded.

2.2 Second Stage: Conceding

After the initial random walk, the agent switches to a conceding strategy. After all, it can hardly make sense to expect something better to come from exclusively random actions than could be achieved with a little more thought. The concession works in a fairly mundane way.

First, all possible bids are enumerated. The bids are then ordered by how much they would detract from the agent's utility and how likely the opponents are inferred to be to accept the bid. That is, at the top of the list come the most advantageous bids that are also the most likely to be accepted, while at the bottom the bids are placed that would be very detrimental to the agent's utility and that would also be unlikely to be accepted by the other agents.

Second, since the list thus constructed is usually going to be too long to concede using every bid in it, tradeoffs will need to be made. The agent keeps track of how long it has taken on average to complete its computation for a single round so that it can estimate how many rounds it has left. (If the negotiation is performed not with a fixed deadline in time but with a fixed number of rounds, that number can be directly compared to the number of the current round.) Once the number of remaining rounds has been determined, the list can be traversed in larger steps so that the concession proceeds as slowly as can be afforded but it will eventually condescend to the other agents, with the caveat that the agent will not concede more than a predefined threshold.

2.3 Opponent Model

To construct the list of bids to concede over, information is required about the utilities that the opponents assign to them. This information is gathered according to the agent's opponent model. The model essentially simplifies the multiparty case to a two-party case by aggregating all the bids it receives from all the opponents and treating them all equally. This is done to avoid the problem of having to determine which of the agents to favor, since this can vary over time and it is not known whether it depends on what the agent itself decides to do (i.e. other agents may start acting more or less favorably depending on what the agent does). The way the likelihood is inferred that a bid will be accepted by the opponents, then, is by determining which values for the issues in the domain occur the most frequently, under the assumption that most agents will not want to change those issue values that they assign the most utility to.

2.4 Final Round

As mentioned above, at the end the agent can perform a special action: if the agent will be the one to make or break the negotiation in the final round, it offers the bid that has the most utility. The reasoning behind this is that the opponents will have to accept the bid since if they do not, the negotiation fails and they are guaranteed to have no better utility than if they accept the agent's unfavorable bid, since in that case every agent's utility will be zero.

3 Experiments and Evaluation

In order to test the proposed strategy, we have run a series of tournaments with three negotiators, i.e. strategies. In this test we have used a single domain with three predefined preference profiles. The way the test was performed was as follows. For each negotiation, the three negotiating agents each get assigned one of the preference profiles. A tournament is then defined as the full set of possible assignments of preference profiles to strategies, so in our case, with three negotiators and as many preference profiles, each tournament had 9 negotiating sessions. This way the success of one strategy cannot be caused by a fortunate set of preferences. For consistency, we have also run each tournament ten times. The average utilities of the agents involved in the tournaments are shown in Table 1. The average utility is calculated as the average utility achieved for each strategy over all 90 sessions.

After running the tournaments, we made the following observations. The strategy we developed was able to secure a reasonable utility most of the time, and when a low utility was achieved, often the opponents did not do much better. The social

Table 1 Average utilities for Pokerface and two other agents in various combinations

Agents	Utility agent 1	Utility agent 2	Utility Pokerface
1,2,10	0.500638	0.666667	0.462303
1,3,10	0.836782	0.806024	0.875527
1,4,10	0.827411	0.792666	0.833574
1,5,10	0.832148	0.834867	0.896057
1,6,10	0.646843	0.744032	0.630657
1,7,10	0.590085	0.726235	0.61285
1,8,10	0.86718	0.833462	0.890016
1,9,10	0.85454	0.837195	0.881035
2,3,10	0.666667	0.500638	0.462303
2,4,10	0.833333	0.598477	0.576198
2,5,10	0.833333	0.598477	0.576198
2,6,10	0	0	0
2,7,10	0	0	0
2,8,10	0.833333	0.598477	0.576198
2,9,10	0.833333	0.598477	0.576198
3,4,10	0.819146	0.824691	0.874389
3,5,10	0.817558	0.832124	0.911351
3,6,10	0.623134	0.670556	0.623728
3,7,10	0.545499	0.663128	0.545698
3,8,10	0.829912	0.850828	0.895889
3,9,10	0.858612	0.815228	0.905506
4,5,10	0.814208	0.790749	0.872897
4,6,10	0.703732	0.895098	0.758256
4,7,10	0.710305	0.908972	0.719138
4,8,10	0.85014	0.739171	0.860858
4,9,10	0.849609	0.836095	0.875939
5,6,10	0.717296	0.862815	0.742641
5,7,10	0.677432	0.90866	0.701149
5,8,10	0.830288	0.773093	0.884417
5,9,10	0.832949	0.8449	0.857725
6,7,10	0.044834	0.048214	0.040297
6,8,10	0.869227	0.709545	0.737495
6,9,10	0.906101	0.770134	0.738663
7,8,10	0.938778	0.732944	0.749235
7,9,10	0.888307	0.701957	0.737068
8,9,10	0.825846	0.841614	0.83903

welfare, however, was clearly a secondary concern. We argue that it is not sensible to make it the primary objective of the agent to try to please everyone, since especially in multiparty negotiation, it soon becomes impossible to find a solution that makes everyone happy, at least in domains that give options that are directly opposed to each other.

4 Conclusions and Future Work

We have described a possible strategy that may be followed in order to enable an automated negotiation agent to negotiate with more than one opponent. The strategy we proposed favors individual utility over social welfare under the (common sense) assumption that it is increasingly difficult to please all opponents when their number increases. The strategy consists of performing random bids for a portion of the available time, then conceding for the remainder of the negotiation session. If at the end no agreement has been reached and we can force the hands of the opponents, we try to do so.

Some possible improvements can be formulated for future work as well. One important feature that is missing at this point is a more detailed opponent model that takes into account the individual opponents' utilities. As outlined above, right now their desires are only considered in aggregate, but it may be possible to improve social welfare by still aligning our concession to the wishes some subset of opponents. It may also be possible to improve the random walk by strategically modifying the threshold according to a model of how likely the current bids are to be accepted by all opponents: if this likelihood is low, it is not dangerous to make a disadvantageous bid, so in that way the randomness may be improved somewhat.

References

1. R. Aydoğan, D. Festen, K.V. Hindriks, C.M. Jonker, Alternating offers protocol for multilateral negotiation, in *Modern Approaches to Agent-based Complex Automated Negotiation*, ed. by K. Fujita, et al. (Springer)
2. T. Baarslag, K. Fujita, E.H. Gerding, K. Hindriks, T. Ito, N.R. Jennings, C. Jonker, S. Kraus, R. Lin, V. Robu, C.R. Williams, Evaluating practical negotiating agents: results and analysis of the 2011 international competition. Artif. Intell. **198**, 73–103 (2013). doi:10.1016/j.artint.2012.09.004
3. T. Baarslag, K. Hindriks, C. Jonker, A tit for tat negotiation strategy for real-time bilateral negotiations, in *Complex Automated Negotiations: Theories, Models, and Software Competitions* (Springer, Heidelberg, 2013), pp. 229–233
4. T. Baarslag, K. Hindriks, C.M. Jonker, S. Kraus, R. Lin, The first automated negotiating agents competition (ANAC 2010), in *New Trends in Agent-based Complex Automated Negotiations*, Studies in Computational Intelligence, ed. by T. Ito, M. Zhang, V. Robu, S. Fatima, T. Matsuo (Springer, Heidelberg, 2012), pp. 113–135. doi:10.1007/978-3-642-24696-8_7

5. K.Y. Gal, L. Ilany, The fourth automated negotiation competition, in *Next Frontier in Agent-based Complex Automated Negotiation*, ed. by K. Fujita, T. Ito, M. Zhang, V. Robu (Springer, Japan, 2015), pp. 129–136. doi:10.1007/978-4-431-55525-4
6. R. Lin, S. Kraus, T. Baarslag, D. Tykhonov, K. Hindriks, C.M. Jonker, Genius: an integrated environment for supporting the design of generic automated negotiators. Comput. Intell. **30**(1), 48–70 (2014). doi:10.1111/j.1467-8640.2012.00463.x
7. C.R. Williams, V. Robu, E.H. Gerding, N.R. Jennings, An overview of the results and insights from the third automated negotiating agents competition (ANAC2012), in *Novel Insights in Agent-based Complex Automated Negotiation*, ed. by I. Marsá-Maestre, M.A. López-Carmona, T. Ito, M. Zhang, Q. Bai, K. Fujita (Springer, Japan, 2014), pp. 151–162. doi:10.1007/978-4-431-54758-7_9

Negotiating with Unknown Opponents Toward Multi-lateral Agreement in Real-Time Domains

Siqi Chen, Jianye Hao, Shuang Zhou and Gerhard Weiss

Abstract Automated negotiation has been gained a mass of attention mainly because of its broad application potential in many fields. This work studies a prominent class of automated negotiations – multi-lateral multi-issue negotiations under real-time constraints, where the negotiation agents are given no prior information about their opponents' preferences over the negotiation outcome space. A novel negotiation approach is proposed that enables an agent to obtain efficient agreements in this challenging multi-lateral negotiations. The proposed approach achieves that goal by, (1) employing sparse pseudo-input Gaussian processes (SPGPs) to model opponents, (2) learning fuzzy opponent preferences to increase the satisfaction of other parties, and (3) adopting an adaptive decision-making mechanism to handle uncertainty in negotiation.

1 Introduction

Negotiation is ubiquitous in our daily life and serves as an important approach to facilitate conflict-resolving and reaching agreements between different parties.

This paper is a shortened version of our previous work [6].

S. Chen (✉)
School of Computer and Information Science, Southwest University, Chongqing, China
e-mail: siqi.chen09@gmail.com

J. Hao
School of Software, Tianjin University, Tianjin, China
e-mail: jianye.hao@tju.edu.cn

S. Zhou · G. Weiss
Department of Knowledge Engineering, Maastricht University, Maastricht, The Netherlands
e-mail: shuang.zhou@maastrichtuniversity.nl

G. Weiss
e-mail: gerhard.weiss@maastrichtuniversity.nl

© Springer International Publishing AG 2017

K. Fujita et al. (eds.), *Modern Approaches to Agent-based Complex Automated Negotiation*, Studies in Computational Intelligence 674,
DOI 10.1007/978-3-319-51563-2_17

Development of automated negotiation techniques enables software agents to perform negotiations on behalf of human negotiators. This can not only significantly alleviate the huge efforts of human negotiators, but also aid human in reaching better negotiation outcomes by compensating for the limited computational abilities of humans when they deal with complex negotiations [13].

During negotiations, an agent usually keeps its strategy and preference as private information, in order to avoid possible exploitation. Thus one major research challenge is to effectively estimate the negotiation partner's preference profile [2, 10, 14, 21] and predicate its decision function [15, 20]. On one hand, through getting a better understanding of partners' preferences, it would increase the possibility of reaching mutually beneficial outcomes. On the other hand, with effective strategy prediction it enables negotiation agents to maximally exploit their negotiating partners and thus receive as much benefit as possible [9]. Until now, fruitful research efforts have been devoted to developing automated negotiation strategies and mechanisms in a variety of negotiation scenarios [4, 7, 13–15, 19]. However, most research efforts have been devoted to bilateral negotiation scenarios, which only models the strategic negotiation among two parties. However, in real life the more common and general way of negotiations usually involve multiple parties. It is in common agreement from the automated negotiation research community that more attention should be given to multilateral negotiations and investigate effective negotiation techniques for multilateral negotiation scenarios.

In this work, a novel negotiation approach is proposed for intelligent agents to negotiate in multilateral multi-issue real-time negotiation. During negotiation, the agents' negotiation strategies and preference profiles are their private information, and the available information about the negotiating partner is its past negotiation moves [12]. Due to the huge strategy space that a negotiating partner can consider, it is usually hard to predict which specific strategy the agent is employing based on the very limited amount of information. Toward this end, instead of predicting the exact negotiation strategies of the opponents, we adaptively adjust the non-exploitation point λ to determine the perfect timing that we should stop further exploits the opponents, and then determine the aspiration level (or the target utility) for proposing offers to opponents before and after the non-exploitation point following different rules. The value of λ is determined as the timing when the estimated expected future utility we can obtain over all opponents is maximized. The future utility that each opponent offers can be efficiently predicted using the Sparse Pesudo-inputs Gaussian Process (SPGP) technique by dividing the negotiation history into a number of atomic intervals.

Given the aspiration level for offering proposals, another important question is how should we select an optimal proposal to reach efficient agreements with other parties, which can also improve the possibility of accepting this offer by the negotiating partners. In this work, we measure the efficient degree of an outcome from a practical perspective – the social welfare of participants. We propose modeling the preferences of each opponent using the least square error regression technique on the basis of negotiation history. After that, the offer with the highest social welfare is selected as the offer to be proposed with certain exploration.

The remainder of the work is organized as follows. Section 2 introduces the multilateral negotiation model given in this work. In Sect. 3, our negotiation approach is then introduced in details. Finally, conclusion and future work are given in Sect. 4.

2 Multilateral Negotiation Model

To govern the complex process of a multilateral negotiation, we adopt an extension of a basic bilateral negotiation protocol [17] which is widely used in the agents field [5, 6, 8, 11, 19]. The participating agents try to establish a contract for a product (service) or reach consensus on certain matter on behalf of their parties [6]. Precisely, let $A = \{a_1, a_2, \ldots, a_i, \ldots, a_m\}$ be the set of negotiating agents, J be the set of issues under negotiation with j a particular issue ($j \in \{1, \ldots, n\}$ where m is the number of issues) [3, 5, 6, 8]. Following the alternating bargaining model of [17], each agent, in turn, has a chance to express its opinion about the ongoing negotiation. The opinion can be communicated in a form of a contract proposal (e.g., a new offer), or an acceptance of the latest offer on the table (note that previous offers would not be accepted once there exists a new proposal), or terminating the negotiation according to its interpretation of the current negotiation situation. A simple illustration of the

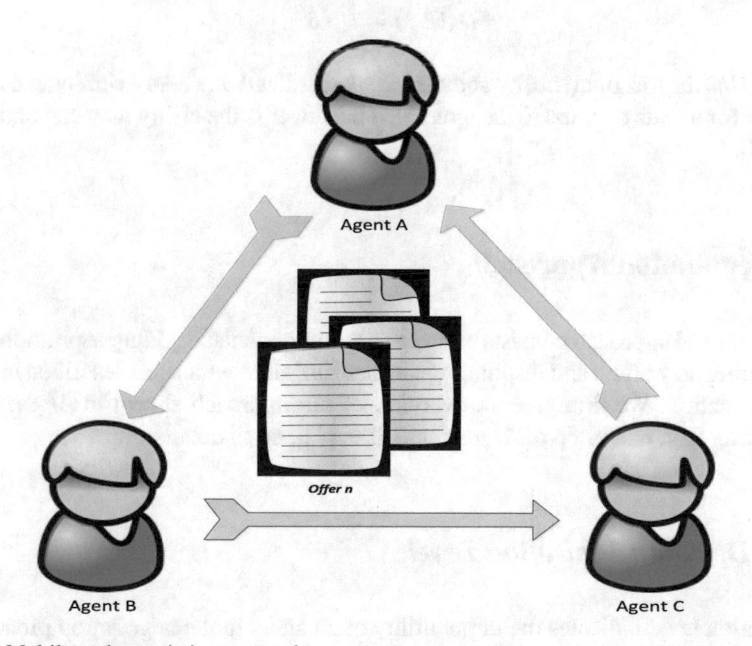

Fig. 1 Multilateral negotiation protocol

multilateral negotiation process is shown in Fig. 1. Due to space constraints we refer the interested reader to the work [1] for further details [3, 5, 6, 8].

An offer is taken as a vector of values, with one value for each issue. The utility of an offer for agent i is calculated by the utility function defined as follows:

$$U^i(O) = \sum_{j=1}^{n} (w^i_j \cdot V^i_j(O_{j,k})) \tag{1}$$

where w^i_j and O are as defined above and V^i_j is the evaluation function of agent i for issue j, mapping every possible value of issue j (i.e., $O_{j,k}$) to a real number [3, 5, 6, 8]. The weight vector **w** denotes the weighting preference of an agent, in which w^i_j represents its preference for issue j. The issue weights of an agent are normalized (i.e., $\sum_{j=1}^{n} w^i_j = 1$ for each agent i). In addition an agent has a lowest expectation for the outcome of a negotiation – the reservation value ϑ [3, 5, 6, 8].

In this work we consider negotiation being conducted in a real-time way. Each negotiator has a hard deadline by when it must have completed or withdraw the negotiation [3, 5, 6, 8]. The negotiation deadline is simply denoted by $t_{\max}$. For domains where the value of agreements is discounted over time, the discounting factor δ ($\delta \in [0, 1]$) is defined to calculate the discounted utility as follows:

$$D(U, t) = U \cdot \delta^t \tag{2}$$

where U is the (original) utility and t is the standardized time. As an effect, the longer it takes for agents to come to an agreement the lower is the utility they can obtain [3, 5, 6, 8].

3 Negotiation Approach

The proposed approach consists of three core components: deciding aspiration level, generating new offers and responding mechanism, all of which are described in detail in this section. We first give an overview of our approach shown in *Algorithm* 1. Following that, each step of *Algorithm* 1 is explained in details.

3.1 Deciding Aspiration Level

Aspiration level indicates the target utility of an agent in the negotiation process. In order to respond to uncertainty in a negotiation where opponents' private information is unknown, the aspiration level is updated due to the environment (e.g., available negotiation time and discounting effect) and opponent behaviors. The agent can therefore predict opponent future moves to assist its decision by analyzing past moves

Algorithm 1 The overview of the proposed negotiation approach. Let t_c be the current time point, δ the time discounting factor, and t_{max} the deadline of negotiation. O_{opp} is the latest opponent offer, Ω_i the previous offers of opponent i and O_{own} a new offer to be proposed by our agent. χ is the time series including the average utilities over intervals. E denotes the expected utility of incoming counter-offers. λ is the non-exploitation time point and u' the target utility. $\mathbf{W}$ denotes the set of learnt opponent weight vectors [3, 5, 6, 8].

1: *Require:* $\vartheta, \delta, t_{max}$
2: **while** $t_c <= t_{max}$ **do**
3: $O_{opp} \Leftarrow receiveMessage;$
4: $\Omega_i \Leftarrow recordOfferSet(t_c, O_{opp}, i);$
5: **if** $myTurn(t_c)$ **then**
6: **if** $updateModel(t_c)$ **then**
7: $\chi \Leftarrow preprocessData(t_c)$
8: $E \Leftarrow Predict(\chi, \Omega);$
9: $(\lambda, U_{\min}) \Leftarrow updateParas(t_c);$
10: $\mathbf{W} = updatePrefreenceModels();$
11: **end if**
12: **end if**
13: $u' = getTargetUtility(t_c, E, \lambda);$
14: $O_{own} \Leftarrow constructOffer(u', \mathbf{W});$
15: **if** $isAcceptable(u'_c, O_{opp}, t_c, \delta)$ **then**
16: $accept(O_{opp});$
17: **else**
18: $checkTermination();$
19: $proposeNewBid(O_{own});$
20: **end if**
21: **end while**

of the opponent. The prediction technique we use here is a computationally efficient variant of standard Gaussian Processes (GPs) – Sparse Pseudo-inputs Gaussian Processes (SPGPs), which proves effective in negotiation context [8]. Another advantage of SPGPs over other type of regression techniques is that it not only provides accurate prediction but also the measure of confidence in the prediction.

Following the notation of GPs in [16], given a data set $\mathscr{D} = \{\mathbf{x}^{(i)}, y^{(i)}\}_{i=1}^{n}$ where $\mathbf{x} \in \mathbb{R}^d$ is the input vector, $y \in \mathbb{R}$ the output vector and m the number of available data points when a function is sampled according to a GP, so we write, $f(\mathbf{x}) \sim \mathscr{GP}(m(\mathbf{x}), k(\mathbf{x}, \mathbf{x}'))$, where $m(\mathbf{x})$ is the mean function and $k(\mathbf{x}, \mathbf{x}')$ the covariance function, fully specifying a GP. Learning in a GP setting involves maximizing the marginal likelihood of Eq. 3 [3, 5, 6, 8].

$$\log p(\mathbf{y}|\mathbf{X}) = -\frac{1}{2}\mathbf{y}^T \left(\mathbf{K} + \sigma_n^2\mathbf{I}\right)^{-1} \mathbf{y} - \frac{1}{2}\log|\mathbf{K} + \sigma_n^2\mathbf{I}| - \frac{n}{2}\log 2\pi, \quad (3)$$

where $\mathbf{y} \in \mathbb{R}^{m\times 1}$ is the vector of all collected outputs, $\mathbf{X} \in \mathbb{R}^{m\times d}$ is the matrix of the data set inputs, and $\mathbf{K} \in \mathbb{R}^{m\times m}$ is the covariance matrix with $|.|$ representing the determinant.

The problem with GPs is that maximizing Eq. 3 is computationally expensive due to the inversion of the covariance matrix $\mathbf{K} \in \mathbb{R}^{n\times n}$ where n is the number of data points. For this specific reason we employ a fast and more efficient learning technique – SPGPs. The most interesting feature of SPGPs is that these approximators are capable of attaining very close accuracy in both learning and prediction to normal GPs with only a fraction of the computation cost. Using only a small amount of pseudo-inputs, SPGPs are capable of attaining very similar fitting and prediction results to normal GPs [3, 5, 6, 8].

When a counter-proposal from agent i arrives at time t_c, our agent records the time stamp t_c and the utility $U(O^i)$ that is evaluated in our agent's utility space. To reduce misinterpretation of the opponent's behavior as much as possible that is caused by the setting of multi-issue negotiations, the whole negotiation is divided into a fixed number (denoted as ζ) of equal intervals. The average utilities at each interval with the corresponding time stamps, are then provided as inputs to the SPGPs. Results in [18] show a complexity reduction in the training cost (i.e., the cost of finding the parameters of the covariance matrix) to $\mathcal{O}(M^2N)$ and in the prediction cost (i.e., prediction on a new set of inputs) to $\mathcal{O}(M^2)$ [3, 5, 6, 8].

After learning a suitable model, SPGPs makes forecast about the future concession of the opponent as shown in line 7 of Algorithm 1. Our agent keeps track of the expected discounted utility based on the predictive distribution at a new input $t_\star$, which is given by:

$$p(u_*|t_\star, \mathcal{D}, \bar{\mathbf{X}}) = \int p(u_\star|t_\star, \bar{\mathbf{X}}, \bar{\mathbf{f}})p(\bar{\mathbf{f}}|\mathcal{D}, \bar{\mathbf{X}})d\bar{\mathbf{f}} = \mathcal{N}(u_\star|\mu_\star, \sigma_*^2), \tag{4}$$

where

$$\mu_\star = \mathbf{k}_\star^T \mathbf{Q}_M^{-1}(\boldsymbol{\Lambda} + \sigma^2\mathbf{I})^{-1}u$$
$$\sigma_\star^2 = \mathbf{K}_{\star\star} - \mathbf{k}_\star^T(\mathbf{K}_M^{-1} - \mathbf{Q}_M^{-1})\mathbf{k}_\star + \sigma^2$$
$$\mathbf{Q}_M = \mathbf{K}_M + \mathbf{K}_{MN}(\boldsymbol{\Lambda} + \sigma^2\mathbf{I})^{-1}\mathbf{K}_{NM}$$

With given probability distribution over future received utilities and the effect of the discounting factor, the expected utility $E_{t_\star}$ is then formulated by

$$E_t = \frac{1}{C}\int_0^1 D(u \cdot p(u; \mu_t, \sigma_t), t)du \tag{5}$$

where $\mu_\star$ and $\sigma_\star$ are the mean and standard deviation at time $t_\star$, and the normalizing constant C is introduced to preserve a valid probability distribution [3, 5, 6, 8].

Our agent employs the target utility function as given in Eq. 6 to determine the aspiration level over time. The function adopts a tough manner (i.e., slowly conceding) before the non-exploitation time point (λ) for seeking higher expected profits, then it quickly goes to the expected minimal utility such that negotiation failure/disagreement could be avoided. λ is tweaked according to the behavior of the negotiation participants. More precisely, the higher the average opponent concession (measured in the our own utility space), the later our agent begins to compromise.

$$u' = \begin{cases} U_{\max} - \Delta\left(\frac{t_c}{\lambda}\right)^{1+\delta} & when\ t_c \leq \lambda, \\ (U_{\max} - \Delta)\left(1 - \frac{t_c - \lambda}{t_{\max} - \lambda}\right)^{1+\delta} & otherwise \end{cases} \tag{6}$$

where $U_{\max}$ is the maximal utility, $U_{\min}$ is the minimal utility ($U_{\min} = \max(\vartheta, \gamma)$ and γ the received lowest opponent concession), constant Δ is the maximal concession amount (i.e., $U_{\max} - U_{\min}$), with

$$\lambda = \operatorname*{argmax}_{t \in T} \frac{1}{|A| - 1} \sum_{i \in A \backslash o} \frac{1}{C_i} \int_0^1 D_\delta(u \cdot p(u; \mu_t, \sigma_t), t) du \tag{7}$$

with o representing our agent and $T \in [t_c, t_{\max}]$.

3.2 Generating Offers

Given an aspiration utility level to achieve, our agent next needs to consider what offer to send such that the likelihood of an offer being accepted could be maximized. Performing this task would require certain knowledge about opponents' preferences. However, negotiation opponents unfortunately have no motivation to reveal their true likings over proposals (or their utility functions) to avoid exploitation. In order to address this problem, we model the opponent concession tactics as time-dependent tactics (originated in [11]) shown in Eq. 8, which are classic tactic in the current literature.

$$\widetilde{u} = U_{\max} - (U_{\max} - \vartheta)(t_c/t_{\max})^{\alpha} \tag{8}$$

where α is the concession factor controlling the style of concessive behavior (for example, boulware behavior ($\alpha < 1$) or conceder behavior ($\alpha > 1$)). Time-dependent tactics are widely used in automated negotiation community to decide concession toward opponents since an negotiator needs to make more or less compromise over time so as to resolve conflicts of the parties. In more detail, boulware tactic maintains the target utility level until the late stage of a negotiation process, whereupon it concedes to the reservation utility. By contrast, conceder tactic makes quick compromise to other parties once a negotiation session starts. For linear tactic, it simply reduces the target utility from the maximal utility to the reservation utility in a linear way.

Learning opponent preferences, while useful, is indeed challenging because information about opponent preferences over different issues (e.g., the weight vector $\mathbf{w}$) is severely lacking. To tackle this issue, researchers typically assume that opponent concession tactic is fully known or preferences follow a certain distribution. In many real-world applications, it is however difficult or costly to acquire the exact information about opponent concession.[1] Therefore we make a mild assumption that we

[1]Note that the opponent concession is the amount of concession measured in the utility space of the opponent instead of ours.

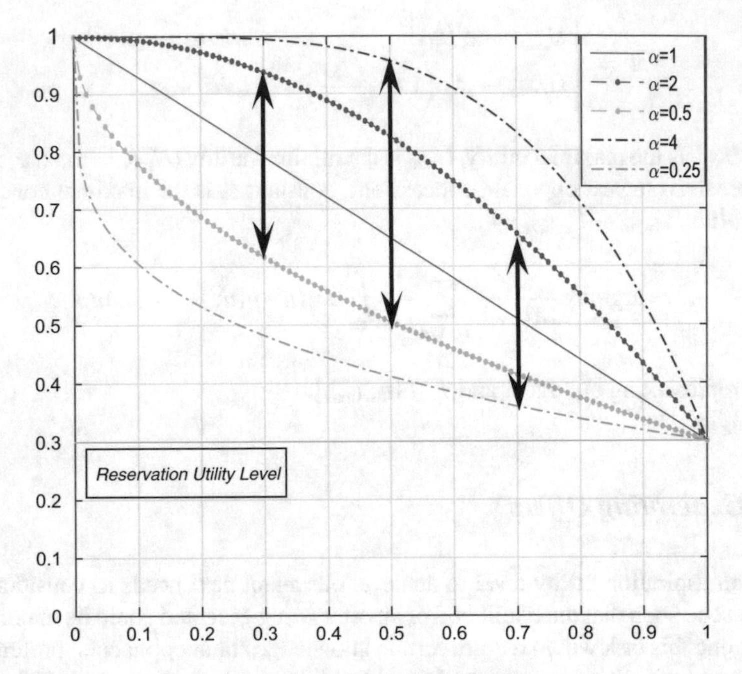

Fig. 2 A toy example of opponent concession ranges given by the pairs of concession factors (0.5, 2) at time 0.3, (0.5, 4) at 0.5 and (2, 0.25) 0.7, respectively

could enquire of domain experts about the approximate concession range of an opponent. This fuzzy knowledge is provided in form of a pair of concession factors that indicate the upper and lower concession an opponent makes at each time point. This idea is illustrated in Fig. 2. Thus, the agent can estimate opponent preferences with the aid of the fuzzy information about opponent concession. Specifically, the preferences are learnt through minimizing the loss function L, which gives the expected loss associated with estimating opponent concession based on a weight vector. The loss function is constructed as in Eq. 10. The loss is calculated by the difference between the mean of concession and the utility of an offer based on a weight vector $\mathbf{w}$; moreover an additional penalty is imposed by φ when an expected utility for $\mathbf{w}$ excesses the upper and lower bounds of opponent concession. When calculating an offer's utility for opponent i, yet the valuation of each issue choice is needed. We here simply assume that the importance order of issue choices is known, and approximate the valuation like [14] as follow,

$$V^i_{j,k}(O_{j,k}) = \frac{2r^i_{j,k}}{K(K+1)} \tag{9}$$

where K is the number of possible choices for issue j, while $r^i_{j,k}$ denotes the ranking of the issue choice $O_{j,k}$.

Let the opponent utility of an offer for a weight vector $\mathbf{w}$ be $\hat{u}_{\mathbf{w}}$. With the opponent concession tactic given in Eq. 8 and the two concession factors (which denote the approximate concession range suggested by experts), our agent can estimate the weight vector of opponent i by means of linear least squares. This can be achieved by minimizing the following loss function,

$$L^i(\mathbf{w}) = \begin{cases} |\frac{(u^i_{upper}+u^i_{lower})}{2} - \hat{u}_\mathbf{w}| + \varphi(u^i_{lower}, \hat{u}_\mathbf{w}), & \hat{u}_\mathbf{w} \leq u^i_{lower} \\ |\frac{(u^i_{upper}+u^i_{lower})}{2} - \hat{u}_\mathbf{w}| + \varphi(\hat{u}_\mathbf{w}, u^i_{upper}), & u^i_{upper} \leq \hat{u}_\mathbf{w} \\ |\frac{(u^i_{upper}+u^i_{lower})}{2} - \hat{u}_\mathbf{w}|, & otherwise \end{cases} \quad (10)$$

with u^i_{upper} and u^i_{lower} being the upper and lower bound of concession made by opponent i at time t, and φ the penalty function as below,

$$\varphi(x, y) = \beta|x - y|^{\frac{1}{2}} \quad (11)$$

where β denotes the confidence of the expert, and the lower the value, the more confidence the expert has about the perdition (to limit further complexness, we let β be 1).

After the estimation of weight vectors of other parties has been done, our agent chooses an offer being capable of maximizing the social welfare (e.g., the sum of the utility of all participants in the negotiation) given a aspiration level, shown as below:

$$\underset{O}{\operatorname{argmax}} \frac{1}{|A| - 1} \sum_{i \in A \backslash o} (\hat{u}^i_\mathbf{w}(O) - \vartheta)^2$$
$$\text{subject to} \quad (12)$$
$$U^o(O) \geq u'$$

Although opponent preferences could be learnt on the basis of the provided concession tactics, it sometimes may be ineffective due to the fuzzy nature of the information; therefore our agent needs an alternative approach to choosing new offers. Fortunately, a real-time negotiation typically allows agents to exchange a large number of offers, thereby giving them many opportunities to explore the outcome space. Therefore, the proposed approach generates a new offer for next round following an ε-greedy strategy. The strategy selects either a greedy action (i.e., exploit) with 1-ε probability ($\varepsilon \in [0, 1]$) or a random action with a probability of ε [3, 5, 6, 8]. It is worth noting that random action means choosing one offer from the set whose utility is above the given aspiration level by chance. The greedy action aims at choosing an offer that are expected to satisfy other sides' preferences most in order to improve their utilities over the negotiation outcome and the chance of the offer being accepted through fuzzy preference learning. With a probability $1 - \varepsilon$, the approach randomly picks one of those offer whose utility is equal or larger than the given aspiration level. In the latter case, the agent just chooses a new offer that has an utility within some range around u'.

3.3 *Responding Mechanism*

This responding mechanism of the proposed approach corresponds to lines 15–20 of Algorithm 1. After receiving a counter-proposal, the agent should decide whether to accept the proposal by checking two conditions. Firstly the agent validates whether the utility of the latest offer from opponents is higher than u', while in the second, the agent needs to determine whether it had already proposed this very offer (i.e., the opponent's counter-offer) earlier. If either one of these two conditions is satisfied, the agent then accepts the offer as shown in line 16 and the negotiation will be completed if the proposal is also supported by the remaining agents [3, 5, 6, 8].

Moreover, when the negotiation situation becomes hard and might offer our agent a utility even lower than the reservation utility, the agent should consider whether to leave the negotiation course to receive the predefined reservation utility or not. Thus the reservation value is taken as an alternative offer from opponents with a constant utility. Thus the agent needs to check if the aspiration utility is smaller than the reservation utility. If positive, our agent is going to leave the negotiation table in the next round. If our agent decides neither to accept the latest counter-proposal nor to leave the negotiation, it proposes a new offer following the steps of lines 19 of Algorithm 1.

4 Conclusion

This work introduced a novel approach for multilateral agent-based negotiation in complex environments (i.e., multiple issues, real time-constrained, and unknown opponents). Our proposed approach, based on the adaptive decision-making scheme and the effective preference learning method, outclassed recent top ANAC agents. Empirical evaluation (see [6]) shows that our agent ont only generates a higher mean individual utility but also leads to better social welfare compared to the state-of-the-art negotiation agents, and further game-theoretic analysis clearly manifests the high robustness of the proposed approach.

Acknowledgements This work is supported by National Natural Science Foundation of China (Program number: 61602391), and also by Southwest University and Fundamental Research Funds for the Central Universities (Grant number: SWU115032, XDJK2016C042).

References

1. R. Aydoğan, D. Festen, K.V. Hindriks, C.M. Jonker, Alternating offers protocols for multilateral negotiation, in *Modern Approaches to Agent-based Complex Automated Negotiation*, ed. by K. Fujita, Q. Bai, T. Ito, M. Zhang, R. Hadfi, F. Ren, R. Aydoğan (Springer, To be published)
2. T. Baarslag, K. Hindriks, C. Jonker, Acceptance conditions in automated negotiation, in *ACAN'11* (2011)

3. S. Chen, H.B. Ammar, K. Tuyls, G. Weiss, Optimizing complex automated negotiation using sparse pseudo-input Gaussian processes, in *Proceedings of the 12th International Joint Conference on Automomous Agents and Multi-Agent Systems* (Saint Paul, Minnesota, ACM, USA, 2013), pp. 707–714
4. S. Chen, H.B. Ammar, K. Tuyls, G. Weiss, Using conditional restricted boltzmann machine for highly competitive negotiation tasks, in *Proceedings of the 23th International Joint Conference on Artificial Intelligence* (AAAI Press, 2013), pp. 69–75
5. S. Chen, J. Hao, G. Weiss, K. Tuyls, H.-F. Leung, Evaluating practical automated negotiation based on spatial evolutionary game theory, in *KI 2014: Advances in Artificial Intelligence*, Lecture Notes in Computer Science, vol. 8736, ed. by C. Lutz, M. Thielscher (Springer, New York, 2014), pp. 147–158
6. S. Chen, J. Hao, G. Weiss, S. Zhou, Z. Zhang, Toward efficient agreements in real-time multilateral agent-based negotiations, in *Proceedings of the 2015 IEEE 27th International Conference on Tools with Artificial Intelligence (ICTAI), ICTAI'15* (IEEE Computer Society, Washington, DC, USA, 2015), pp. 896–903
7. S. Chen, G. Weiss, An efficient and adaptive approach to negotiation in complex environments, in *Proceedings of the 20th European Conference on Artificial Intelligence*, vol. 242 (IOS Press, Montpellier, France, 2012), pp. 228–233
8. S. Chen, G. Weiss, An intelligent agent for bilateral negotiation with unknown opponents in continuous-time domains. ACM Trans. Auton. Adapt. Syst. **9**(3), 16:1–16:24 (2014)
9. S. Chen, G. Weiss, An approach to complex agent-based negotiations via effectively modeling unknown opponents. Expert Syst. Appl. **42**(5), 2287–2304 (2015)
10. R.M. Coehoorn, N.R. Jennings, Learning an opponent's preferences to make effective multi-issue negotiation trade-offs, in *ICEC'04* (2004), pp. 59–68
11. P. Faratin, C. Sierra, N.R. Jennings, Negotiation decision functions for autonomous agents. Rob. Autom. Syst. **24**(4), 159–182 (1998)
12. J. Hao, H.-F. Leung, ABiNeS: an adaptive bilateral negotiating strategy over multiple items, in *Proceedings of the 2012 IEEE/WIC/ACM International Conference on Intelligent Agent Technology* (IEEE Computer Society, Macau, China, 2012), pp. 95–102
13. J. Hao, S. Song, H.-F. Leung, Z. Ming, An efficient and robust negotiating strategy in bilateral negotiations over multiple items. Eng. Appl. Artif. Intell. **34**, 45–57 (2014)
14. K. Hindriks, D. Tykhonov, Opponent modeling in auomated multi-issue negotiation using bayesian learning, in *AAMAS'08* (2008), pp. 331–338
15. B. Jakub, K. Ryszard, Predicting partner's behaviour in agent negotiation, in *AAMAS '06* (2006), pp. 355–361
16. C.E. Rasmussen, C.K.I. Williams, *Gaussian Processes for Machine Learning* (The MIT Press, Cambridge, 2006)
17. A. Rubinstein, Perfect equilibrium in a bargaining model. Econometrica **50**(1), 97–109 (1982)
18. E. Snelson, Z. Ghahramani, Sparse gaussian processes using pseudo-inputs, *Advances in Neural Information Processing Systems* (MIT press, Cambridge, 2006), pp. 1257–1264
19. C.R. Williams, V. Robu, E.H. Gerding, N.R. Jennings, Negotiating concurrently with unkown opponents in complex, real-time domains, in *ECAI'12*, pp. 834–839 (2012)
20. D. Zeng, K. Sycara, Bayesian learning in negotiation, in AAAI Symposium on Adaptation, Co-evolution and Learning in Multiagent Systems, pp. 99–104 (1996)
21. D. Zeng, K. Sycara, Bayesian learning in negotiation. Int. J. Hum. Comput. Syst. **48**, 125–141 (1998)

Jonny Black: A Mediating Approach to Multilateral Negotiations

Osman Yucel, Jon Hoffman and Sandip Sen

Abstract We describe the strategy of our negotiating agent, "Jonny Black", which received the 3rd place in ANAC 2015 competition in the "Nash Product" category. The agent tries to act as a mediator to find the best outcome for all the agents, including itself, in the negotiation scenario. The agent models other agents in the negotiation, and attempts to find the set of outcomes which are likely to be accepted by the other agents, and then picks the best offer from its own viewpoint. We give an overview of how to implement such a strategy and discuss its merits in the context of closed multilateral negotiation.

1 Introduction

This paper presents our agent called "Jonny Black" and its strategy, which we developed and entered into the Fifth Automated Negotiating Agent Competition[1] (ANAC2015). ANAC is a tournament between a set of negotiating agents which perform closed multilateral negotiation using the alternating offers protocol. The negotiation environment consists of multi-issue scenarios and includes uncertainty about the opponents preferences.

Bazerman et al. showed that introducing a mediator agent significantly reduces the probability of an impasse [2]. Based on this observation, we chose to design

[1] http://mmi.tudelft.nl/anac.

O. Yucel (✉) · J. Hoffman · S. Sen
The University of Tulsa, Tulsa, OK, USA
e-mail: osman-yucel@utulsa.edu

J. Hoffman
e-mail: hoffman.jon.m@gmail.com

S. Sen
e-mail: sandip-sen@utulsa.edu

© Springer International Publishing AG 2017

K. Fujita et al. (eds.), *Modern Approaches to Agent-based Complex Automated Negotiation*, Studies in Computational Intelligence 674,
DOI 10.1007/978-3-319-51563-2_18

our agent to act as a mediator between the other agents.[2] In the process, it attempts to offer the best outcome for itself from the set of solutions in the intersection of acceptable set of outcomes for both opponents.

We treat the negotiation process as a search process where we try to find an existing outcome within the acceptable region for both opponents. In order to be able to do that we need to know what offers are acceptable for the opponents, which requires a model of the opponent of the form studied in the literature [1, 3, 4].

2 Strategy

We present an outline of our agent's strategy in four steps. First we explain the parameters we used for our strategies. Then we explain how we model the opponent. In the following subsection we explain how our agent decides to accept an offer or not. Finally we explain the strategy out agent uses for choosing the bid to offer to it's opponents.

2.1 Parameters

The agent uses 5 parameters to make decisions: *Minimum Offer Threshold (MoT), Agreement Value (AV), Care (C), Number Of Bids to Consider From Opponent's Best Bids (N), and Reluctance (R)*. We will now explain what these parameters are used for and how they are calculated.

2.1.1 Minimum Offer Threshold

Minimum Offer Threshold (MOT) is a constant value used to eliminate the bids from consideration which give our agent less than a desirable outcome. We have empirically set this value to 0.6. With this setting our agent will never offer a bid which gives itself a utility which is less than 0.6.

2.1.2 Agreement Value

Agreement Value (AV) is a parameter that is recalculated with each offer, which determines if our agent should accept an offer or not. The calculation of this variable will be explained in Sect. 2.4.

[2]Every round of negotiation takes place between 3 agents.

2.1.3 Care

Care (C) is a parameter that represents how much we care about our opponent's happiness. Our agent assumes that the opponent will accept a bid which gives the opponent a utility greater than the care value. Care value starts with the value 0.4 and increased by 0.4% every 10 turns. Both of these values are experimentally determined. Even though the negotiations in the competition were multilateral, we only consider one of the opponents at a time. We will explain how the multilateral negotiation aspect is addressed in Sect. 2.5.

2.1.4 Number of Bids to Consider from Opponent's Best Bids

Number Of Bids to Consider From Opponent's Best Bids (N) is a parameter we use for the calculation of the Agreement Value. This parameter determines how many of the best bids we created will be considered for each opponent. As mentioned in Sect. 1, we are trying to offer within the acceptable range of the opponents and this parameter allows us to predict the range that we believe the opponent will accept. Initially we keep this range wide, and narrow it down over time. We start this value from 100 and decrease it by 5 every 10 turns until it gets down to 10, after which it is held steady at 10. The use of this parameter will be explained in Sect. 2.4.

2.1.5 Reluctance

Reluctance (R) is the parameter we use for tracking time. It makes our agent less willing to accept the bids offered during the early periods of the negotiation while becoming more willing with the passage of time. We use this parameter while calculating the Agreement Value. The value of Reluctance starts at 1.1 and is decreased by 0.5% every 10 turns.

2.2 Initialization

During the initialization of the agent, we calculate every possible bid which gives us a greater utility than MOT. We refer to this set of bids as $Bids_{Feasible}$.

2.3 Opponent Modeling

In this section, we will explain the working of our opponent modeling algorithm. Our modeling approach tries to find the opponent's weights on the issues and their relative preferences of the options for every issue. We have two main assumptions in the opponent modeling process:

- The more preferred option for the opponent appears more frequently in the bids offered or accepted by that opponent.
- The opponent will be less likely to change from its best option for more important issues.

2.3.1 Order of Options

To find the preference order and the value of options for an opponent, we calculate the frequency of the options' appearance in the bids accepted and proposed by the opponent. When we order the options for an issue by their frequencies, we do so using V_o, the value predicted for option o. It is calculated using n_o, the rank of the option o, and k, the number of possible options for the issue:

$$V_o = \frac{k - n_o + 1}{k} \tag{1}$$

This calculation values the most frequent option as 1 and least frequent option as $\frac{1}{k}$.

2.3.2 Weights of Issues

As mentioned above, we are assuming that the opponents will be less willing to change the option for more important issues. We use *Gini Index* [5] as the impurity measure, and the issues which have higher Gini-Impurity scores are weighted more by the opponent. $\hat{w}_i$ is the unnormalized weight of issue i, f_o is the frequency of option o, and t is the total number of prior bids as follows

$$\hat{w}_i = \sum_{o \in O_i} \frac{{f_o}^2}{t^2}, \text{where } O_i \text{is the set of options for issue } i \tag{2}$$

Using $\hat{w}_i$, we calculate w_i, the normalized weight of issue i as:

$$w_i = \frac{\hat{w}_i}{\sum_{j \in I} \hat{w}_j}, \text{ where I is the set of issues} \tag{3}$$

2.3.3 Example Model

We now provide an example scenario to illustrate our modeling approach. The frequencies of offered or accepted options in the issues by the opponents for our example is presented in Table 1.

Table 1 The frequencies of options for issues for one opponent

	Option 1	Option 2	Option 3
Issue 1	9	1	0
Issue 2	3	5	2

Table 2 The calculated values for every option

	Option 1	Option 2	Option 3
Issue 1	1	0.66	0.33
Issue 2	0.66	1	0.33

The order of the option preferences are the same as the order of the frequencies of the options. So the relative order in the issues are $I_1(O_1 > O_2 > O_3)$ and $I_2(O_2 > O_1 > O_3)$. From the formula given in Sect. 2.3.1 we calculate the values for the options as $V(I_1, O_1) = \frac{3-1+1}{3} = 1$. We calculate the values for all the issue-option pairs using the same approach. The results of the calculations are listed in Table 2.

We then calculate the weight for the issues using the formula given in Sect. 2.3.2

$$\hat{w}_1 = \frac{9}{10}^2 + \frac{1}{10}^2 = \frac{82}{100}$$
$$\hat{w}_2 = \frac{3}{10}^2 + \frac{5}{10}^2 + \frac{2}{10}^2 = \frac{38}{100} \tag{4}$$

We proceed by normalizing the values we calculated

$$w_1 = \frac{\frac{82}{100}}{\frac{82}{100} + \frac{38}{100}} = \frac{82}{120}$$
$$w_2 = \frac{\frac{38}{100}}{\frac{82}{100} + \frac{38}{100}} = \frac{38}{120} \tag{5}$$

For an offer which has the option O_1 for I_1 and the option O_3 for I_2 we predict the valuation for the modeled user to be

$$V(O_1, O_3) = \frac{82}{120} \times 1 + \frac{38}{120} \times 0.33 \approx 0.789 \tag{6}$$

2.4 Accepting Strategy

Every time the agent receives an offer, it checks if the utility of the offer is greater than our parameter AV. The agent accepts only if that condition is met.

Every 10 turns, the following steps are taken to recalculate AV:

- Evaluate every bid in $Bids_{Feasible}$ using the model of each user.
- Find Set_u: the set which contains the N best bids for opponent u.
- Find $Set_{common} = \bigcap_{u \in Opp} Set_u$, where Opp is the set of opponents.
- Find $Bid_{Best} = \underset{b \in Set_{common}}{\text{argmax}} \; Utility(b)$.
- Finally $AV = Utility(Bid_{Best}) * R$.

This approach estimates AV is the best utility we can get by staying in the acceptable region of both ourselves and our opponents. Multiplying it with R prevents the agent from accepting too early in the negotiations or waiting too long.

2.5 *Bidding Strategy*

In this part we explain how out algorithm chooses which bid to make. During the initialization of the agent we sort the bids in $Bid_{Feasible}$ by their utility for our agent, and set the variable $lastBid$ to 1.

Beginning at the previous bid, search $Bid_{Feasible}$ for the next bid that both our agent and the agent we favor will likely accept. If we find that bid, offer it, set $lastBid$ to that bid's index, and switch the agent we favor. If we do not find an acceptable bid in the set between lastBid and the end of $Bid_{Feasible}$, we simply bid the best bid for us, set $lastBid$ to 1 and switch the agent we favor.

Algorithm 8 pseudocode for selection of a bid

```
1: for i = lastBid + 1 to BidSet.size() do
2:     if Util(BidSet[i]) ≥ AgreeVal and
3: Pred.Util(BidSet[i], agentToFavor) ≥ Care then
4:         Offer(BidSet[i]);
5:         lastBid = i; agentToFavor = otheragent
6:         return;
7: Offer(BidSet[1]);
8: lastBid = 1;
9: agentToFavor = otheragent
```

This algorithm goes down the list of bids until a suitable bid cannot be found and then returns to the top of the list. Figure 1 shows the utility of the offers on the table for the participating agents. In this figure, Party 1 represents our agent while Party 2 and Party 3 are instances of Boulware and Conceder agents respectively, which were

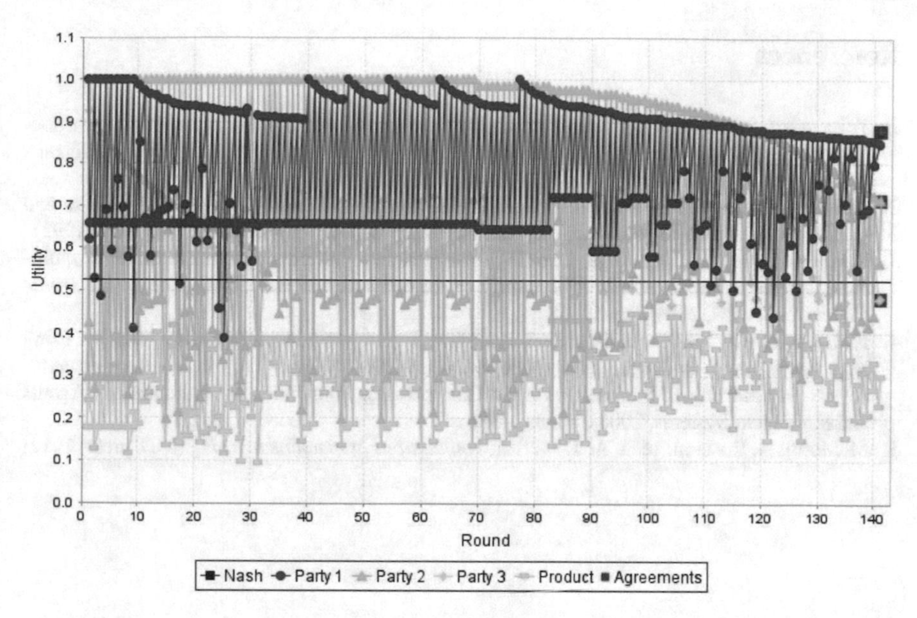

Fig. 1 Utility of the bids for 3 agents in the negotiation

provided by the organizers. The method we mentioned, going down the list as long as the bids are suitable and then returning to top, can be seen from the offers made by Party 1.

3 Conclusion and Future Work

In this paper, we have provided an overview of the strategy of the agent named "Jonny Black", which received the 3rd place in the ANAC 2015 competition in the "Nash Product" category. We have designed our agent to act as a mediator between all participating agents, including and favoring itself. We described the methods used for modeling the opponents and the strategies for accepting and making offers.

The results of the tournament shows that our approach was successful at increasing the social welfare, which was our initial intent while deciding the strategy. However the results also revealed that our agent may concede too much from its own payoff while it is trying to increase the social welfare. For further improvements on this agent, the method of conceding on its own utility for achieving higher social utility should be revised to restrict the agent from conceding too much.

Acknowledgements We would like to thank Tandy School of Computer Science in The University of Tulsa, for making Chapman Distinguished PhD-Student award, which made the focus on this study possible.

References

1. T. Baarslag, M. Hendrikx, K. Hindriks, C. Jonker, Measuring the performance of online opponent models in automated bilateral negotiation, in *AI 2012: Advances in Artificial Intelligence* (Springer, 2012), pp. 1–14
2. M.H. Bazerman, M.A. Neale, K.L. Valley, E.J. Zajac, Y.M. Kim, The effect of agents and mediators on negotiation outcomes. Organ. Behav. Hum. Decis. Process. **53**(1), 55–73 (1992)
3. R.M. Coehoorn, N.R. Jennings, Learning on opponent's preferences to make effective multi-issue negotiation trade-offs, in *Proceedings of the 6th International Conference on Electronic Commerce* (ACM, 2004), pp. 59–68
4. K. Hindriks, D. Tykhonov, Opponent modelling in automated multi-issue negotiation using bayesian learning, in *Proceedings of the 7th International Joint Conference on Autonomous Agents and Multiagent Systems-Volume 1* (International Foundation for Autonomous Agents and Multiagent Systems, 2008), pp. 331–338
5. M. Mohri, A. Rostamizadeh, A. Talwalkar, *Foundations of machine learning* (MIT press, 2012)

Agent X

Wen Gu and Takayuki Ito

Abstract Agent X is an agent which is designed for ANAC2015—The Sixth International Automated Negotiating Agents Competition. In this paper, we will show the two main strategies—Agreement Behaviour and Conceder Behaviour of Agent X. Although a lot of agents pay more attention to the individual utility, Agent X is an agent which thinks highly of the social welfare. As a result, Agent X tries its best to come to an agreement with other agents before the negotiation ends. But it doesn't mean that Agent X doesn't care the result of its individual utility, Agent X also tries to get a relatively good individual utility at the same time.

1 Introduction

As we want to design an Agent which may be practically useful in real world in the future, we make our Agent prioritize social welfare instead of individual utility. Because we think that it will be much more meaningful to the society if the three agents get respectively 0.6, 0.7, 0.7 as their individual utility, other than get respectively 0.9, 0.5, 0.5. In the rules of ANAC2015 [1], there is a time limit and a discount factor in about half of the domains,where the value of an agreement decreases over time. As a result, we consider that it is very important to successfully come to an agreement with other agents and also make it as soon as we can. Based on the those understandings of ANAC2015 [1], we design the strategy of Agent X.

W. Gu (✉) · T. Ito
Nagoya Institute of Technology, Nagoya, Japan
e-mail: gu.wen@itolab.nitech.ac.jp; koku.bun@itolab.nitech.ac.jp

T. Ito
e-mail: ito.takayuki@nitech.ac.jp

© Springer International Publishing AG 2017

K. Fujita et al. (eds.), *Modern Approaches to Agent-based Complex Automated Negotiation*, Studies in Computational Intelligence 674,
DOI 10.1007/978-3-319-51563-2_19

2 Strategy of Agent X

There are two main parts in Agent X's strategy—Agreement Behaviour and Conceder Behaviour.

When acting the agreement behaviour, Agent X tries to make partial agreement with one agent before trying to come to an agreement with both other agents. When acting the conceder behaviour, Agent X tries to successfully come to an agreement in the end and tries to make it as soon as possible.

2.1 Agreement Behaviour

In ANAC2015 [1], the focus of its competition is on multi-party negotiation. We think that it will be difficult if we want to come to an agreement with both of other agents at the same time. As a result, we make Agent X try to make partial agreement first. Just as shown in Fig. 1, our agent tries to have a partial agreement with the agent where we get bids from.

We estimate the number of the bids that Agent A may agree with by calculating the average of Agent A's recent 10 bids. But if Agent A is an agent which is very easy to act a conceder behaviour, we can not have a satisfying social welfare. In order to avoid this kind of situation happen, we import a restraining factor—if the average of Agent A's 10 bids is smaller than 0.618, we will offer the bid by ourself.

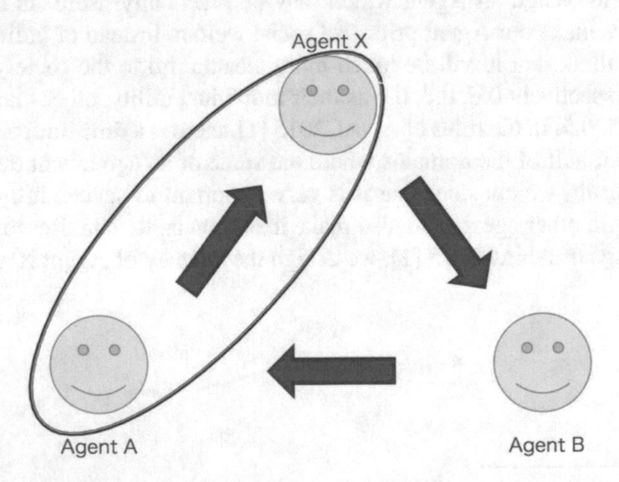

Fig. 1 Try to make partial agreement with the agent where Agent X get bids from

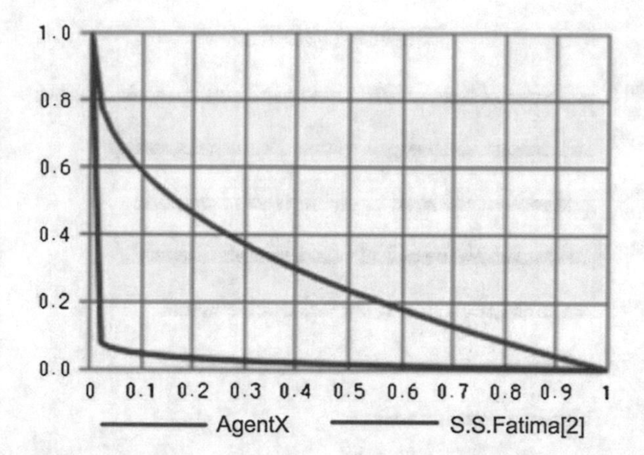

Fig. 2 Using S.S.Fatima formula as a reference to design our conceder behaviour

2.2 *Conceder Behaviour*

In ANAC2015 [1], one of the noteworthy rules of encounter is that there is a time limit during the negotiation, and there is a discount factor which makes the value of the entire negotiation decrease over time. As an agent which thinks highly of the social welfare, agent X tries to come to an agreement as soon as possible. That's why we make our agent easy to act conceder behaviour. Referring to the formula by Faratin [2], we design our agent's strategy. As shown in Fig. 2, we make some modification to let our become more and more easy to act conceder behaviour as the time goes by. We also try to be hard-headed at the beginning of the negotiation in order to get an relatively good individual utility and avoid the situation I refer in the Agreement Behaviour part.

3 Results of Simulation

We did a simulation by using all the agents we designed in our laboratory. Agent X worked very well in achieving satisfying social welfare just as shown in Fig. 3. At the same time, Agent X's average individual utility is 0.75217 which is also relatively satisfying. In addition, Agent X is also the agent which use least time to come to an agreement with other agents.

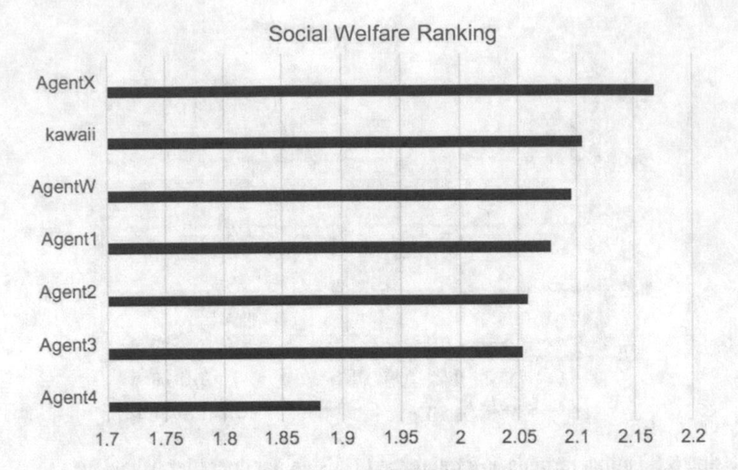

Fig. 3 The results of the social welfare in the simulation we did in our laboratory

4 Conclusion

Based on our understanding of ANAC2015 [1], we design Agent X as an agent which thinks highly of social welfare. We focus on how to come to an agreement with other agents before the negotiation ends and try to use time as less as possible. At the same time, we do not ignore the individual utility. We also try to get an satisfying individual utility in the condition of being able to have a good social welfare.

References

1. The Sixth International Automated Negotiation Agents Competition (ANAC2015) http://www.tuat.ac.jp/~katfuji/ANAC2015/
2. P. Faratin, C. Sierra, N.R. Jennings, Negotiation decision functions for autonomous agents. Robot. Auton. Syst. **24**(3), 159–182 (1998)

CUHKAgent2015: An Adaptive Negotiation Strategy in Multilateral Scenario

Chi Wing Ng, Hoi Tang Leung and Ho-fung Leung

Abstract Automated negotiation has been an active research topic. Many previous works try to model opponents' behaviours using some mathematical functions. However, this approach turns out not to be very effective due to inaccessibility of information and dynamics of the negotiation environment. In this paper, we introduce an adaptive strategy for automated negotiation. We propose that only the concession degree of the opponent is estimated in the negotiation. We also use an adaptive non-exploitation point on the timeline of negotiation for stopping making concession to the opponent. Finally, we suggest that the bid that we propose should be as close to the Pareto Frontier of the bids as possible. The agent CUHKAgent2015 that is based on this strategy showed good performance in ANAC 2015.

Keywords CUHKAgent2015 · Adaption · Pareto Frontier

1 Introduction

Negotiation is an important element of human society, as conflicts are usually solved by negotiation. Automated negotiation can help us to save time and efforts from tiring negotiation as computers have much higher computational power.

Many diversified algorithms and strategies have been implemented in automated negotiation. The main challenge for automated negotiation is the incompleteness

---

C.W. Ng
Department of Physics, The Chinese University of Hong Kong, Sha Tin, Hong Kong, China
e-mail: 1155034174@link.cuhk.edu.hk

H.T. Leung
Department of Information Engineering, The Chinese University of Hong Kong,
Sha Tin, Hong Kong, China
e-mail: 1155032495@link.cuhk.edu.hk

H.-f. Leung (✉)
Department of Computer Science and Engineering, The Chinese University of Hong Kong,
Sha Tin, Hong Kong, China
e-mail: lhf@cuhk.edu.hk

© Springer International Publishing AG 2017
K. Fujita et al. (eds.), *Modern Approaches to Agent-based Complex Automated Negotiation*, Studies in Computational Intelligence 674,
DOI 10.1007/978-3-319-51563-2_20

of information. During negotiation, the preference profiles and decision algorithms of the opponents are kept unknown, in order to protect the opponents from being exploited. So the difficulty becomes how to model the opponents' behaviour and predict the opponent's decision, so that we can make our decisions accordingly.

Previous work usually assumes that opponent's behaviour can be accurately modeled in the form of some mathematical functions, which turns out to be ineffective due to the highly dynamic negotiation environment, thus we propose that only the concession degree of the opponent is estimated in the negotiation. This can also achieve the purpose of modeling opponent's behaviour to a certain extent, while saving much of computing time to make fast response to the opponent.

We also use the non-exploitation point λ on the timeline of negotiation to stop making too much concession to the opponent [4]. λ can be changed during the negotiation, so that we can make adaptive response to the opponent.

In terms of bid-proposing, we suggest that the bid that we propose should be as close to the Pareto Frontier of the bids as possible. This ensures the highest utility for the opponent while choosing bids from the same utility level of us. A scoring algorithm is used in CUHKAgent2015 to achieve this purpose.

The remainder of this paper is structured as follows. In Sect. 2, we describe some important issues on designing our negotiation strategy. The negotiation strategy of CUHKAgent2015 is explained in detail in Sect. 3. Finally, we draw conclusions in Sect. 4.

2 Design Issues

In this section, we discuss 3 key issues on designing an effective negotiation strategy for CUHKAgent2015.

2.1 Reasons of Not Predicting Opponent's Decision Function

Recent efforts have focused on how to learn and predict the opponent's decision function, in order to predict the opponent's next offer [2, 3, 5]. In this approach, the opponent's strategy is estimated by various mathematical models based on different assumptions. For example, the concession curve of the opponent can be calculated based on the assumption that the opponent will compromise more at a later time [5]. Some studies [2, 3] attempt to use a standard statistical modeling, for instance Gaussian processes (GPs), to predict the opponent's decision function.

If the opponent's decision function can be predicted accurately, this helps the agent to identify the optimal exploitation so that the agent can maximise its own utility obtained in the negotiation. However, due to the high uncertainty of opponent's behaviour, it is hard to predict the opponent's decision function accurately. The difference between the predicted and actual opponent's decision functions becomes large when the agent faces some unexpected and complicated behaviour of the opponent. Besides that, the approaches to predict the opponent's decision function are ineffective due to the limited negotiation time and computational power. Even there exists a perfect model which takes into consideration all the relevant bidding history, the actual decision function used by the opponent may be changed during the negotiation, thus the accuracy of long-term prediction cannot be guaranteed.

Hence, instead of studying how to estimate opponent's decision function, we decide to model the opponent's response at a macroscopic level, i.e. the opponent's concession degree. Then, our agent can make decisions adaptively. Detailed discussion will be made in Sect. 3.2.

2.2 *Factors Determining Degree of Concession to Opponent*

During the negotiation, the concession degree to the opponent is adjusted adaptively based on 3 factors, namely the discounting factor δ, the time left for further negotiation t, and the concession degree of the opponent.

The first one is related to the urgency for our agent to achieve an agreement. In a negotiation domain with small discounting factor, a low utility will result if the negotiation time is too long. This also raises the importance of the second factor. As the time left for negotiation gets shorter, the utility of a given bid is lower due to the discounting factor. As a result, we cannot only focus on how to obtain the highest possible utility by being tough, but also how to reduce the utility loss due to the discounting factor. The third factor is about how much utility we should exploit the opponent so that we can obtain the highest possible utility. If the opponent is concessive, then our agent can also try to be more concessive, which hopefully can lead to a win–win agreement.

2.3 *Guessing Opponents' Preferences - Approaching the Pareto Frontier*

In many cases, there are many bids on the same utility level with respect to our agent, then it becomes important on how to propose a bid that is the most favourable to the opponents. This can maximise the chance of the bid being accepted by the opponent. As a result, we try to propose bids as close to the Pareto Frontier as possible.

As the exact preference profile of the opponent is hard to determine, we can instead select a reference bid R according to the opponent's previous bids. Then we can calculate the Euclidean distance between any candidate bid B and R in the issue space of the domain, which acts as an estimate for the value of B to the opponent. Further discussion will be made in Sect. 3.4.

3 Strategy Description

In this section, we describe how CUHKAgent2015 applies the negotiating principles above to the actual negotiation process.

3.1 Overall Flow of Decision

CUHKAgent2015 divides the whole decision making process into several components: Threshold-determining, Bid-proposing and Final Decision. When the agent receives offers from opponents, the offers will be recorded in corresponding bidding history of the proposing opponent. When it comes to CUHKAgent2015 to decide an action to take, a threshold for acceptance and bidding is computed for each opponent by the Threshold-determining component. The Bid-proposing component will then combine the different thresholds for all opponents into one and propose a bid according to the combined threshold. The Final Decision component will compare the bid with the last bid received and reservation value of the domain to reach an action that benefits CUHKAgent2015 more.

3.2 Separated Acceptance and Bidding Thresholds for Different Opponents

For CUHKAgent2015, two different thresholds will be determined by the threshold-determining component: acceptance threshold u^{acc} and bidding threshold u^{bid} for each of the opponents. Any bid with its utility higher than its acceptance threshold u^{acc} can be accepted. Similarly, our agent should only offer to the opponent any bid with its utility higher than bidding threshold u^{bid}. The reason why we use different thresholds for different proposes is described later in this section.

We first describe how to analyse the characteristics of the negotiation domain to set our minimum acceptance threshold u^{min}. There are 4 factors in the domain which affects u^{min}, namely discounting factor δ, reservation value without discounting u^{res}, the average utility of bids u^{avg} and the corresponding standard deviation σ of utilities of bids in the domain without discounting.

For discounting factor δ, the smaller δ is, the lower the utility of the bids become when time goes by. To arrive at an agreement earlier, especially if δ is small, a lower minimum acceptance threshold is needed so that we can accept less favourable bids. Meanwhile, in a worst-case scenario, an agreement may never be reached, hence no agent can get any utility at the end if $u^{res} = 0$. As a result, other than δ, we still need to compromise more to reduce the risk of reaching no agreement when u^{res} is smaller. In our strategy, we use $u^{avg} + \sigma$ as the targeted value for u^{min}. However, this target can be lowered if δ and u^{res} are small. In any case, this value should not be lower than u^{res}. Mathematically, u^{min} is calculated as follows:

$$u^{min} = max\left(u^{res}, u^{avg} + \sigma \times \delta \times u^{res}\right)$$

In general, during the negotiation, we face with 2 basic types of opponent's responses: cooperative and competitive ones. Cooperative opponents seek to create a win–win situation, while competitive ones tend to maximise their own utility only, which might lead to a win–lose outcome. Our strategy is that we make concessions first to offer utility gains to the opponent in order to get a friendly response. If we get friendly responses from the opponents continuously, we expect that eventually we can reach a win–win agreement with high utility for us. However, the more competitive the opponent is, the more utility we need to give up to reach an agreement. Making continuous concessions to the opponent causes a low utility result to us. Hence, we use the adaptive non-exploitation point λ, which is a dynamic time deadline specifying the time when we should not continue making further concession. At its core, the value of non-exploitation point depends on the negotiation environment, including the discounting factor and the concession degree of the opponent. The value of the non-exploitation point is changed dynamically by the following two criteria:

- A smaller discounting factor puts time pressure on us to reach an agreement more quickly because the time that can be taken for reaching an agreement is shortened if we want to obtain a utility as close to the original utility of the agreement as we can. Thus, if the discounting factor is smaller, the non-exploitation point should also be moved earlier.
- On the other hand, to figure out the opponent's concession degree, our agent divides the bids offered by the opponents into clusters of fixed number of consecutive bids. For each cluster, the bid with the highest utility based on our own preference profile is selected as a characteristic bid. We assume that the opponent who provides a sequence of characteristic bids with increasing utility is willing to offer a win–win outcome. The higher the utility of the characteristic bid in the latest cluster compared with that of the characteristic bid in the previous cluster, the more concessive the opponent we consider. Specifically, the opponent's concession degree γ_t is defined as follows,

$$\gamma_t = (u^{latest} - u^{previous})/\Delta t$$

where γ_t is the estimated concessive degree of the opponent at current time, u^{latest} is the utility of the characteristic bid in the latest cluster, $u^{previous}$ is utility of the characteristic bid in the previous cluster and Δt is the difference of time between the latest cluster and the previous cluster.

In the early stage of negotiation, we cannot collect enough information to determine the opponent's concession degree, so the value of non-exploitation point will be initialised to be a constant. Afterwards, the value of non-exploitation point begins to change over time according to the concession degree of the opponent and the discounting factor. It is determined as follows,

$$\lambda = e^{\delta - 1 + \gamma_t / k}$$

where k is the weighting factor used to enlarge γ_t if the negotiation domain size is small. In the negotiation with small domain size, the variety of the bids the opponent can offer is limited. The bidding of the opponent may not reflect the actual concession degree of the opponent. Thus, we use the weighting factor k to enlarge the concession degree in this kind of negotiation domain.

After we get the adaptive non-exploitation point λ and the minimum acceptance threshold u^{min}, the remaining issue we focus on is how to apply them to our acceptance threshold and bidding threshold.

Since we attempt to search for higher chance of reaching an agreement, a further compromise is made when time $t \leq \lambda$. Thus, the value of bidding threshold decreases with an increase in time and approaches to u^{min} until $t > \lambda$. If $t > \lambda$, we assume that the opponent is likely to seek for a win–lose bid as an agreement, because we have put much effort into compromising to reach a win–win situation before time λ. Thus we should not make further compromise. In this respect, the way the agent should behave is to raise the bidding threshold to put pressure on the opponent to offer better bids for us. Hence, the value of our bidding threshold approaches to u^{max} after time λ. Specifically, the bidding threshold over time is described in detail as follows,

$$\begin{aligned} u^{bid} &= u^{min} + (u^{max} - u^{min})(1 - sin(\tfrac{\pi t}{2\lambda})), \text{ if } t \leq \lambda \\ u^{bid} &= u^{min} + (u^{max} - u^{min})/((1 - \lambda)(t - \lambda)^{\delta}), \text{ if } t > \lambda \end{aligned}$$

where u^{max} is the maximum utility in the negotiation domain without discounting.

Considering that the opponent may select the best bid we ever proposed as one of the opponent's offers [1], we separate the bidding threshold and the acceptance threshold and select the lowest bidding threshold u^{bid} as the acceptance threshold u^{acc}, in order to minimise the risk of reaching no agreement. The overview of acceptance threshold u^{acc} during the negotiation is defined as follows,

$$u^{acc} = u^{bid}, \text{ if } u^{bid} < u^{acc}.$$

3.3 *Combining the Thresholds and Bid-Proposing*

After computing the individual thresholds for the two opponents (u_1^{bid} and u_2^{bid}), the next step is to combine the thresholds into one (u^{bid}), so that the agent can propose a bid according to this threshold. In CUHKAgent2015, the thresholds are simply averaged between the two opponents to get a balance between the interests of the opponents. In the bid-proposing process, the whole set of bids is divided into

20–50 intervals according to the utility of the bids with respect to our agent. Bids are proposed by choosing the best one within the interval that u^{bid} is located in. In order to have a higher probability that the opponents accept our bid, we add a normally distributed probability centering at u^{bid} for the agent to choose a utility threshold higher than the previously determined u^{bid}.

3.4 Approaching the Pareto Frontier

Within an interval, there are often many bids to choose from, which we call them candidate bids. To maximise the benefit of our agent, the best approach is to get as close as possible to the Pareto Frontier, which maximises the opponents' utility within a given interval of bid utility with respect to our agent. While maximising the opponent's utility, the chance of the proposed bid to be accepted by the opponent is also maximised. This also gives, hopefully, the highest chance that the opponent accepts a bid with high utility to us.

In CUHKAgent2015, this is done by building a reference bid R for each of the opponents. The whole negotiation domain consists of a set of issues to be negotiated. Each bid in the domain is formed by various values of all the issues in the domain. R is built by choosing the most frequently proposed value of each issue by the opponent. Then for any candidate bid B, a score S is calculated using the following formula:

$$S = \sum_{i \in I} (R_i - B_i)^2$$

where i is an issue in the negotiation domain, I is the set of all issues, R_i is the value of issue i of the reference bid R, B_i is the value of issue i of the candidate bid B.

This is essentially the square of Euclidean distance of the two bids in the issue space. After the above process, two scores are calculated for each candidate bid, the two scores are then added up and compared with the scores of other candidate bids. As the bid with the lowest score should have the smallest total distance to the two reference bids for the two opponents, it should be the candidate with the highest total utility for the two opponents, i.e. the closest to the Pareto Frontier, so this bid is chosen to be proposed.

4 Conclusion

We propose in this paper an adaptive strategy for automated negotiation. In the strategy, decisions are based on the estimated concession degree of the opponents and an adaptively changing non-exploitation point. Bids to be offered are then selected as close as possible to the Pareto frontier. The agent CUHKAgent2015, which made use of the strategy, performed well in the ANAC 2015.

References

1. S. Chen, G. Weiss, OMAC: a discrete wavelet transformation based negotiation agent, in *Novel Insights in Agent-based Complex Automated Negotiation*, ed. by I. Marsa-Maestre, M.A. Lopez-Carmona, T. Ito, M.J. Zhang, Q. Bai, K. Fujita (Springer, Tokyo, 2014), pp. 187–196
2. S. Chen, H.B. Ammar, K. Tuyls, C. Weiss, Optimizing complex automated negotiation using sparse pseudo-input gaussian processes, in *Proceedings of the 12th International Conference on Autonomous Agents and Multiagent Systems* (2013), pp. 707–714
3. S. Chen, G. Weiss, An approach to complex agent-based negotiations via effectively modeling unknown opponents. Expert Syst. Appl. 2287–2304 (2015)
4. J. Hao, H.F. Leung, CUHKAgent: an adaptive negotiation strategy for bilateral negotiations over multiple items, in *Novel Insights in Agent-based Complex Automated Negotiation*, ed. by I. Marsa-Maestre, M.A. Lopez-Carmona, T. Ito, M.J. Zhang, Q. Bai, K. Fujita (Springer, Tokyo, 2014), pp. 171–179
5. C.R. Williams, V. Robu, E.H. Gerding, N.R. Jennings, IAMhaggler: a negotiation agent for complex environments, in *New Trends in Agent-Based Complex Automated Negotiations*, ed. by T. Ito, M.J. Zhang, V. Robu, S. Fatima, T. Matsuo (Springer, Heidelberg, 2012), pp. 151–158

AgentH

Masayuki Hayashi and Takayuki Ito

Abstract The Automated Negotiating Agents Competition (ANAC2015) was organized. Automated agents negotiate with either other agents or people, trying to make an agreement that maximizes the utility of all of the negotiators. This agent-based negotiation will relieve some of the effort people have during negotiations. Thereby, development in automated agents is an important issue. In the ANAC2015 competition, the fundamental rule is changed; from negotiations between two agents to negotiations among multiple agents. In accordance to this change, we developed AgentH, aiming at successful negotiations in the competition. In this paper, we present the strategy of the agent and the way it searches for new bids to offer, based on the opponents' bids.

1 Introduction

The sixth international Automated Negotiating Agents Competition (ANAC2015) was held [1]. A variety of agents with different strategies have been proposed in ANAC. It is likely that those strategies are be applicable to the real-life negotiation problems.

The negotiations had been done between two agents in the competition until ANAC2014. In the ANAC2015, however, a new rule was introduced [2] and the negotiations were done among more than three agents. In accordance to this change, we needed to modify the existing strategies or construct new strategies so that the agents can negotiate with more than two opponents at the same time. We developed a negotiating agent, AgentH, that has a simple strategy and uses a simple way to search for bids to offer during the negotiation.

M. Hayashi (✉)
Department of Computer Science, Nagoya Institute of Technology, Nagoya, Japan
e-mail: hayashi.masayuki@itolab.nitech.ac.jp

T. Ito
Master of Techno-Business Administration, Nagoya Institute of Technology, Nagoya, Japan
e-mail: ito.takayuki@itolab.nitech.ac.jp

© Springer International Publishing AG 2017

K. Fujita et al. (eds.), *Modern Approaches to Agent-based Complex Automated Negotiation*, Studies in Computational Intelligence 674,
DOI 10.1007/978-3-319-51563-2_21

2 Implementation of AgentH

In this section, we describe the strategy of AgentH and the way the agent searches for new bids to offer.

2.1 *Compromising Strategy*

In real-life negotiations, negotiators first offer the bids that are beneficial for them, whose utilities are relatively high for each of them, and gradually change the way they offer their bids to compromised bids, whose utilities to be less and less, considering the opponents' behavior. Aiming at reproducing this behavior of negotiators in our agent, we designed the agent to have two different behaviors; Hardheaded Behavior and Compromising Behavior, and gradually change the behavior between them.

The Hardheaded Behavior represents the behavior where the agent is arrogant and does not accept the opponents' bids if they are not beneficial for him, but instead offer another bid that are more beneficial for him. In other words, if the utility value of the opponents' bids are relatively low, the agent does not accept them, but instead offers another bid whose utility value is higher than them.

On the other hand, the Compromising Behavior is the behavior where the agent tend to accept non-beneficial bids as well. In this behavior, the agent accepts the opponents' bids even if their utility values are relatively low.

To gradually change the agent's behavior depending on time, we used the following compromising utility function.

$$cu(\boldsymbol{x}, t) = u(\boldsymbol{x}) \times t \tag{1}$$

where $\boldsymbol{x}$ is the bid, t is the normalized time ranging from 0 to 1, and $u(\boldsymbol{x})$ is the utility function. When a bid $\boldsymbol{x}$ is offered by one of the opponents, the agent determines if he should accept it or not by the output of this compromising utility function; the agent accepts the bid if its compromising utility value is higher than a certain threshold, and the agent rejects the bid if its compromised utility value is lower than the threshold. This threshold is set heuristically so that the agent behaves like a human negotiator compromising as the time passes. This time, it is set to 0.45, through some experimental negotiations that we made when making this agent.

2.2 *Offering New Bids*

When the agent does not accept the opponents' bids, it searches for another bid to offer and then offer it. AgentH uses a simple heuristic that the bids which are similar to those offered or accepted by the opponents are likely to be accepted by them as

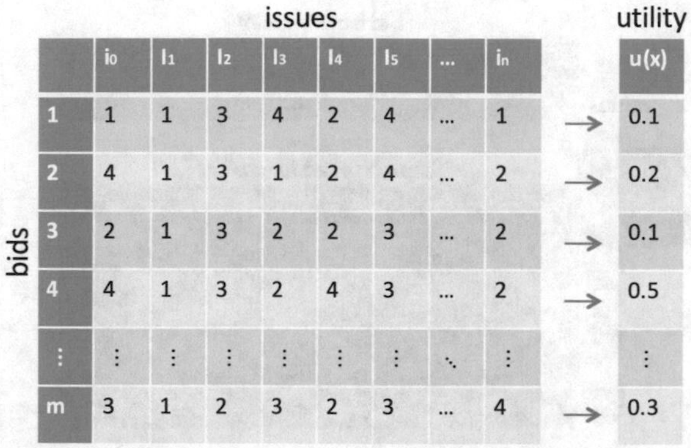

bids	issues i_0	I_1	I_2	I_3	I_4	I_5	...	i_n		utility u(x)
1	1	1	3	4	2	4	...	1	→	0.1
2	4	1	3	1	2	4	...	2	→	0.2
3	2	1	3	2	2	3	...	2	→	0.1
4	4	1	3	2	4	3	...	2	→	0.5
⋮	⋮	⋮	⋮	⋮	⋮	⋮	⋱	⋮		⋮
m	3	1	2	3	2	3	...	4	→	0.3

Fig. 1 Example: an example state of the stored bid history. Each row represents a bid consisting of values assigned as shown by each column

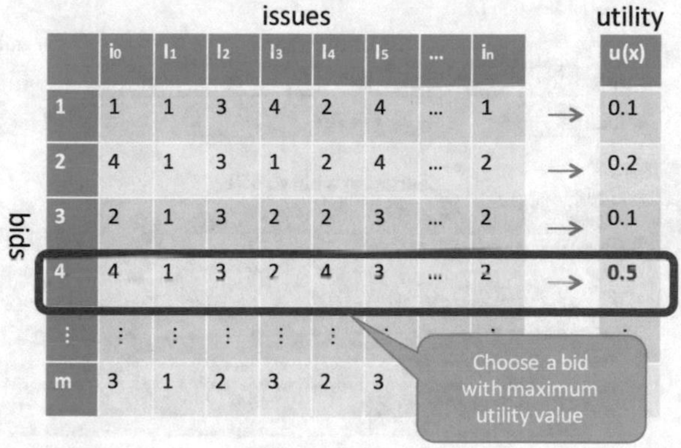

bids	issues i_0	I_1	I_2	I_3	I_4	I_5	...	i_n		utility u(x)
1	1	1	3	4	2	4	...	1	→	0.1
2	4	1	3	1	2	4	...	2	→	0.2
3	2	1	3	2	2	3	...	2	→	0.1
4	4	1	3	2	4	3	...	2	→	0.5
⋮	⋮	⋮	⋮	⋮	⋮	[illegible]	[illegible]	[illegible]		[illegible]
m	3	1	2	3	2	3	[illegible]	[illegible]		[illegible]

Fig. 2 Example: picking up a bid to modify. In this example, the forth bid is chosen as the bid to modify since it has the highest utility value

well. Based on this idea, the agent can offer a bid that are likely to be accepted by the opponents. At the same time, we want the agent to offer bids whose utility value is relatively higher than the bids that are already offered. To achieve this goal, our agent searches for the bid which is slightly different from the existing bids offered or accepted by others, and whose utility value is higher than them. This can be done by modifying one of the existing bids offered or accepted by opponents, so that its utility value becomes higher. We will describe more detail about the methodology.

The whole methodology of searching for the bid is splitted into two procedures: (1) to pick up a bid to modify and (2) to modify the bid so that its utility value becomes higher.

Bid from history

issues	i_0	i_1	i_2	i_3	i_4	i_5	...	i_n		utility u(x)
values	4	1	3	2	4	3	...	2	→	0.5

Randomly choose the value

search for a bid to offer

i_0	i_1	i_2	i_3	i_4	i_5	...	i_n		u(x)
3	1	3	2	4	3	...	2	→	0.3
4	2	3	2	4	3	...	2	→	0.4
4	1	1	2	4	3	...	2	→	0.4
⋮	⋮	⋮	⋮	⋮	⋮		⋮		⋮
4	2	3	2	4	4	...	2	→	0.6

Fig. 3 Example: modifying the bid using the methodology described in Sect. 2.2.2. Values for each issues in the modified bid is changed iteratively to a random value, and then its utility value is calculated

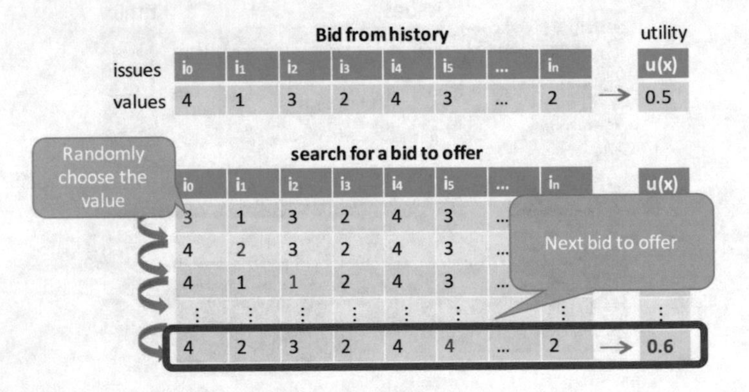

Fig. 4 Example: choosing the bid to offer in the way described in Sect. 2.2.2. The bid whose utility value is higher than the original bid is chosen as the next bid to offer. In this example, the bid at the bottom of the table is chosen with its utility value, 0.6, higher than that of the original bid, 0.5

2.2.1 Picking Up a Bid to Modify

Searching for a bid is done when the agent does not accept the existing bid (denied bid) offered by one of the opponents. When picking up a bid to modify, our agent chooses the bid that was offered or accepted by the opponent who offered the denied bid, especially the one with the maximum utility value for our agent. For example, suppose, for one of the opponents, our agent stores the history of the bids offered or accepted by him as shown in Fig. 1. When our agent deny the last bid (the bid at the bottom of the table in Fig. 1) offered by him, our agent consult the opponent's bid history, and then chooses the fourth bid (see Fig. 2) as the bid to modify because it has the maximum utility value of all the bids offered or accepted by the opponent.

2.2.2 Modifying the Bid

After picking up a bid to modify, our agent modifies the bid so that its utility value becomes higher than the existing bids. As bids are composed of values for each issues, we used a very simple method to modify the bid; iteratively change one of the values for issues to a random value. See Figs. 3 and 4 for an example of this modification. In this way, our agent can find a bid that is barely different but has higher utility value.

3 Conclusion

In this paper, we showed the compromising strategy and the bidding methodology of AgentH. AgentH is an automated negotiating agent that is applicable for negotiating with multiple agents. We showed how its behavior changes by time, and how it searches for new bids that are likely to be accepted by other agents as well as having higher utility values.

References

1. The Sixth International Automated Negotiating Agents Competition (ANAC2015), http://www.tuat.ac.jp/~katfuji/ANAC2015/
2. R. Aydoğan, D. Festen, K.V. Hindriks, C.M. Jonker, Alternating offers protocols for multilateral negotiation, in *Modern Approaches to Agent-based Complex Automated Negotiation*, ed. by K. Fujita, Q. Bai, T. Ito, M. Zhang, R. Hadfi, F. Ren, R. Aydoğan (Springer, To be published)

Printed in the United States
By Bookmasters